T0319289

WOOD COATINGS: THEORY AND PRACTICE

WOOD COATINGS: THEORY AND PRACTICE

FRANCO BULIAN

JON A. GRAYSTONE

ELSEVIER

Amsterdam • Boston • Heidelberg • London • New York • Oxford
Paris • San Diego • San Francisco • Singapore • Sydney • Tokyo

Elsevier
Radarweg 29, PO Box 211, 1000 AE Amsterdam, The Netherlands
The Boulevard, Langford Lane, Kidlington, Oxford OX5 1GB, UK

First edition 2009

© Copyright 2009 Franco Bulian and Jon A. Graystone. Published by Elsevier B.V.
All rights reserved.

Notice
No responsibility is assumed by the publisher for any injury and/or damage to
persons or property as a matter of products liability, negligence or otherwise, or
from any use or operation of any methods, products, instructions or ideas
contained in the material herein. Because of rapid advances in the medical sciences,
in particular, independent verification of diagnoses and drug dosages should be
made

British Library Cataloguing in Publication Data
A catalogue record for this book is available from the British Library

Library of Congress Cataloging-in-Publication Data
A catalog record for this book is available from the Library of Congress

ISBN: 9780444528407

For information on all Elsevier publications
visit our web site at *elsevierdirect.com*

Printed and bound in Hungary
09 10 11 12 10 9 8 7 6 5 4 3 2 1

Working together to grow
libraries in developing countries

www.elsevier.com | www.bookaid.org | www.sabre.org

ELSEVIER   BOOK AID
             International   Sabre Foundation

CONTENTS

Chapter 9 Operational Aspects of Wood Coatings: Application and Surface Preparation   259

This book addresses the factors responsible for the appearance and performance of wood coatings in both domestic and industrial situations. The term 'wood coatings' covers a broad range of products including stains, clearcoats, topcoats and supporting ancillary products that may be used indoors or outdoors. Major market sectors for wood coatings include furniture, joinery and flooring.

Techniques for coating wood go back many centuries but in recent decades there has been a move towards more environmental-friendly materials, for example, the use of water-borne rather than solvent-borne binders. Alongside this has been a growing awareness of 'sustainability' to moderate the environmental impacts of economic growth. As a renewable resource, wood has a special attraction as a material substrate which should be complemented by any necessary coatings.

A major objective of 'wood coatings' is to explain the underlying factors that influence selection and development of coatings within the constraints of the major wood market sectors. This requires integrating the appearance and performance needs within an operational framework. Basic information on the chemistry and technology of coatings is included for the benefit of students and laboratory technicians. Additionally, the book includes individual chapters of interest to architects, specifiers and industrial users.

Jon Graystone is a Principal Research Scientist at PRA (part of PERA Innovation) with 50 years experience in the Coatings Industry including 37 years with ICI Paints (now Akzo Nobel). He has broad experience in Research and Development. His current interests include service life prediction and formulation techniques. He is involved with many of PRA's technical training courses including Paint Technology and Formulation. He has been an active member of TC139/WG2 (Exterior Wood Coatings) since the inception of the committee in 1989 (email: jon@graystone.demon.co.uk).

Franco Bulian is the Vice-Director and Head of the Chemical Department of CATAS, the biggest Italian research institute in the wood and furniture sector. He has acquired a broad experience in wood coatings achieved also by the several research projects he took part in. He is the coordinator of the Italian standardisation committee on wood coatings (UNICHIM) and member of TC139/WG2 (Exterior Wood Coatings). Since 2007 he is a professor at the University of Trieste (email: franco.bulian@libero.it).

Markets for Wood and Wood Coatings

1. PROLOGUE

The total global coatings market volume is estimated at 26.5 billion litres worth in the region of 55 billion Euro [1] with many different types of product. Why does a user choose a particular type of coating system? There are many answers to this question including economic, environmental and operational ones. However, an important overarching factor is the nature of the *substrate* to be coated. Coatings for metal, masonry and plastic would be expected to have some very different requirements, but also perhaps some properties in common. *Wood*, and its derived products, represents another distinct substrate category with many properties that will influence the ideal choice of coating. Wood differs from the other generic substrate categories, in that it is derived from a living organism,

Wood Coatings: Theory and Practice
DOI: 10.1016/B978-0-444-52840-7.00001-1

namely a 'tree'. Not only is it a sustainable resource but also wood has an inherent beauty, which is even imitated in some synthetic materials. One consequence of the aesthetic appeal of wood is the high proportion of transparent finishes and semi-transparent finishes that are found in the marketplace that are designed to enhance the colour and grain of wood.

The fact that wood coatings range from low-build penetrating stains to high-build opaque paint systems raises another interesting question. What is it that they have in common? Surprisingly, the answer is very little! Wood coatings can only be assessed in the context of a specific user-driven need, covering both appearance and operational factors. Different solutions to the problems raised may present themselves, and many types of technology will be applicable albeit with a different balance of properties. End uses for wood coatings range from the domestic 'decorative' market to industrial applications. The versatility of wood finds use in buildings inside and out, in flooring and in furniture. In addition to the many species of tree that provide wood, there is substantial use of various 'composite' forms such as plywood and fibreboard.

In preparing this book, the authors have tried to cover a wide range of interests ranging from students and formulators, to architects, specifiers and manufacturers. Accordingly, most chapters can be treated as modules and do not need to be read in a particular sequence. An overview of the structure of the book is shown schematically in Fig. 1.

2. MARKETS FOR WOOD AND WOOD COATINGS

2.1. Consumption of timber

The world's forests produce large amounts of timber, a valuable and renewable natural resource. For centuries, the pressure on this resource has resulted in a diminution of the great natural forests and an increase in plantation trees. The latter now surpass supplies from natural forests with most resources in the Southern Hemisphere. Bio-engineering has resulted in faster growing species which may be available for sawing in as little as 15 years. Timber grown for pulp and conversion to board products may be grown in even shorter periods. Timber thus remains an important strategic resource and has in recent years attracted interest from the perspective of the carbon content in relation to climate change. This is a complex issue, trees can certainly immobilise carbon, but this may be released later as carbon dioxide during decay processes. It is also significant that around 50% of timber is still used as fuel!

There are many species of trees with many important botanical differences that are described later. A broad distinction may be made between hardwoods and softwoods. Production of hardwoods to softwoods is in

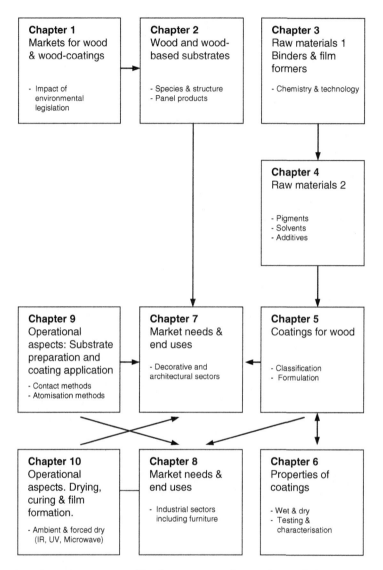

Chapter 1
Markets for wood
& wood-coatings

- Impact of
  environmental
  legislation

Chapter 2
Wood and wood-
based substrates

- Species & structure
- Panel products

Chapter 3
Raw materials 1
Binders & film
formers

- Chemistry & technology

Chapter 4
Raw materials 2

- Pigments
- Solvents
- Additives

Chapter 9
Operational
aspects: Substrate
preparation and
coating application

- Contact methods
- Atomisation methods

Chapter 7
Market needs &
end uses

- Decorative and
  architectural sectors

Chapter 5
Coatings for wood

- Classification
- Formulation

Chapter 10
Operational
aspects. Drying,
curing & film
formation.

- Ambient & forced dry
  (IR, UV, Microwave)

Chapter 8
Market needs &
end uses

- Industrial sectors
  including furniture

Chapter 6
Properties of
coatings

- Wet & dry
- Testing &
  characterisation

Figure 1 Schematic overview of book structure and main chapter links.

the ratio of around 60:40 but this may be changing as some forests become exhausted. To a first approximation, 70% of the hardwoods come from tropical countries and the remainder from temperate climates. The latter areas produce over 90% of softwoods.

Wood materials, particularly lower grade softwoods, are often used in the form of panels. Chipboard is still the major panel product; followed by plywood, however medium density fibreboard (MDF) usage is increasing fairly rapidly at a growth rate around 10%. The MDF market in Europe was reported as €6.8 million in 1999 [2] but this rate of growth has now slowed.

Among the major applications for wood are construction and furniture. A major percentage (~43%) of softwood is used constructing the fabric of buildings with a further 9% used in joinery. Relatively little softwood is used in furniture where hardwood is generally more suitable. Interior doors and windows account for nearly 75% of component manufacture.

2.2. End-use sectors – Coated wood

Wood has many diverse uses, some of which require coating for protective or decorative purposes. End-use sectors are usefully assessed in categories that can be related to the types of coating required. For example, meaningful distinction can be made between interior and exterior applications, the latter being much more demanding in terms of resistance to moisture and solar radiation. Another important distinction is between 'decorative' and 'industrial' applications.

The term 'decorative' is widely used to denote coatings designed for buildings, known in some countries as 'Architectural Coatings'. Such coatings account for more than half of all coating production. The decorative sector may be split into a Professional (Trade) and Retail (DIY) sector. The distinction is not always sharp and can vary from 60:40 to 30:70 according to country [3]. Decorative coatings are applied by hand and are air-drying. This is in contrast to the industrial sector, which uses a wider range of application techniques and products that may require stoving or radiation curing. In the building application, there is overlap between decorative and industrial end-use categories, in that products originally coated in a factory, for example, windows may require subsequent maintenance in situ. As is described in Chapter 7, maintenance painting has a different set of operational requirements to industrial coating, and this will strongly influence the technology used.

2.3. Joinery and windows

The term 'joinery' is used fairly loosely to denote products from the carpenter's craft, and also to denote constructional wood products in building including windows, doors and skirting boards. Wooden window frames are an important sub-sector with special coating requirements in both decorative and industrial sectors.

UPVC (unplasticised polyvinyl chloride) has made major inroads into the market for wooden windows and now dominates the replacement market, which itself represents 65% of total demand. There are several reasons for this, not least the poor design and detailing of some traditional

Table 1 Windows production in Europe

Material	Percentage
UPVC	42
Aluminium	28
Wood–aluminium	6
Wood	24

wooden windows, but also the problems of maintenance. Europe currently produces around 70 million window units [4] of which the materials distribution is approximately shown in Table 1.

There are significant differences between countries; aluminium is most popular in Spain, Portugal and Southern Italy. Wood and wood–aluminium combinations are dominant in Northern Europe, while central Europe and the UK use the most uPVC.

The decline in wooden windows represents a significant loss of market for wood coatings but hostility to PVC from environmental pressure groups may reverse the trend. Several local housing authorities are developing strategies to switch away from PVC, although timber frames are 20% more expensive as well as requiring costly maintenance in high-rise blocks. Greenpeace and other pressure groups are actively seeking to influence the double-glazing window market away from PVC, to timber certified by the Forest Stewardship Council (FSC).

In addition to toxicity, recycling is also an issue connected with uPVC. Hardly, any uPVC window profiles have ever been recycled in the UK, and even in Germany recycling has reached only 7000 tonnes per year. Failure by the industry to develop a suitable recycling infrastructure might result in producer responsibility legislation imposed by the European Commission.

2.4. Furniture

Within Europe, there are estimated to be 90,000 furniture producers, a market worth €84 billion and employing around one million people [5]. A complication of the industry is a disproportion between large and small companies. Big companies (>100 employees) account for 70% of the overall sales and the remainder is supplied by the many SMEs, representing two thirds of the total workforce. Smaller companies are at a particular disadvantage when adopting some new technologies and therefore particularly under pressure from impending legislation as described later.

The total market structure is complex covering domestic, office and contract sectors. Many materials and methods are used including soft furnishings and upholstery. Wood may be hidden as part of a structural supporting frame, but is also very visible in some sectors where its appearance is valued. Wood is also a component of many panel products and may be used in the form of a veneer.

As well as responding to legislation, the selection of coatings for wood furniture involves operational factors such as whether it is supplied fully assembled, or as flat stock ('knock-down'). Within Europe, there are regional and national preferences for the appearance of the finished product. Southern Europe has a preference for more glossy finishes, while in the North a more 'open-pore' effect may be preferred. Individual species of wood can present unique problems as exemplified by oak, which has large open pores and presents a difficult surface to achieve smooth coatings.

2.5. Industrial wood coatings

Statistics on coating markets are always difficult to come by, and this is particularly true of wood coatings where there are many statistical anomalies. Part of the difficulty lies in the definitions used. Certainly, some companies use the term 'woodcare' for what might loosely be called stains (lasures) and varnishes, whereas more traditional paint systems may be put in a different category even though a substantial amount is used on wood. Moreover, wood coatings straddle the industrial and decorative sectors of joinery, but may be dealt with as 'industrial finishes' when describing furniture coatings. Flooring presents another difficulty since it may be coated industrially or in situ. There is also a growing sector of products used outdoors but more as part of garden rather than house decor. These will include coatings for decking, and brighter non-traditional colours used on fences for decorative rather than protective purposes. It should also be noted that volume can be expressed in different units such as weight, volume or monetary value in absolute or relative (percentage) terms. This may obscure growth trends as new sectors are defined, or as higher solids replace lower or indeed as margins are squeezed.

World paint output currently stands at around 26.5 billion litres with the 10 largest companies taking 43% of the market. The European share of this is about 5.8 million tonnes split 55:45 between decorative and industrial sectors.

The European industrial paint breakdown indicates wood finishes to have a sizeable share. Figures from CEPE [6] and elsewhere indicate this sector (for 13 countries) to be around 400,000 tonnes.

In terms of individual countries, Italy has the largest share reflecting a major position in furniture production. However, it should be noted that these figures do not include Eastern and Central Europe which could add a further 255,000 tonnes.

Within the broad definition of industrial wood finishing, furniture at ~56% is by far the largest share, outstripping joinery at ~30%. The remainder includes flooring such as parquet, which is enjoying a growth phase.

Overall Italy, at 140,000 tonnes, has the largest industrial wood-coatings market in Europe, but Germany accounts for the largest share of furniture production: 26% as against Italy's 22%, in spite of the general movement of the furniture industry from Germany to Eastern Europe. The Italian furniture industry, however, employs the most people: 224,000, compared to 216,000 in Germany.

There are some interesting differences between the take up of technologies between countries, which reflect operational differences, and national preferences.

2.6. Parquet and wood flooring

In global terms, wood has a relatively small share of the market between 4% and 5%, compared with ceramic tiles (46%) and carpets (33%) [7]. Nonetheless, this still represents a significant floor area and one that is growing within Europe. Oak is the dominant wood species accounting for nearly half of the volume, followed by beech and tropical hardwoods. The total floor area is estimated at 84 mm^2 [8]. Parquet itself, however, is under pressure from laminate flooring which holds a similar share of the market.

Parquet can be coated in a factory finishing process; though much will be coated in situ, there then also arises the problem of maintenance. Cleary, a major requirement for flooring is a very tough, durable scratch-resistant clear coating with a first coat which must enhance appearance ('Anfeuerung') and provide good adhesion.

2.7. Major players in industrial wood coatings

The major players in Europe are Akzo Nobel, Becker Acroma, Milesi and Hickson all of whom have followed acquisition strategies.

Consolidation has swept through the European furniture industry in recent years, but German producers still dominate the European top 10 league: Schieder, Steinhoff, Welle, Alno and Wellman. Only one Italian furniture maker, Natuzzi, can claim membership in that league but there is strong cooperation between smaller Italian enterprises which act as sub-contractors to larger ones.

There is also growing consolidation among the manufacturers of furniture finishes, though national concerns are still strong, especially in Germany, which have largely remained in German hands. Apart from BASF and Becker Acroma (Swedish), they include Hesse Lignal,

3H Lacke and Votteler. In Italy, in contrast, market leadership belongs to the Italian Milesi Group, but acquisitions have introduced foreign multi-nationals into the industry, including BASF, Arch Chemicals and Akzo Nobel.

2.8. The decorative 'woodcare' market

In many ways, it is even more difficult to pin down market sizes for decorative than industrial wood-coating markets, differences in terminology make for fluid definitions. The overall market for DIY ('do it yourself') paints is around 0.8 billion litres with the trade sector even bigger at ~1.7 billion. Against these large markets, stain and varnishes are often classed as 'niche' sectors though they are bigger than many industrial sectors [9].

2.8.1. Varnishes
The market is claimed to be around 60 million litres with Akzo Nobel, BASF, Dyrup, Ronseal and Williams as the main players.

2.8.2. Wood stains
Wood stains are similar in size to varnishes at 55 million litres, but this probably underplays the overlap with paints. Major players are the same as varnishes though as a result of acquisitions the brand names are different. Akzo include now Sadolin and Sikkens, ICI (now part of Akzo) own Cuprinol, Dyrup cover Bondex, Desowag, Gori, Xylol, while Williams include Consolan and Xylamon.

2.9. Technology breakdown

The wood-coating sector uses many different types of resin and polymer under such headings as 'nitrocellulose', 'alkyd', 'acrylic', 'polyester', 'polyurethane', etc. The chemistry and properties of these different materials are described in later chapters. However, chemical terminology does not describe the technological factors that are important and cover the means by which a particular chemistry is delivered, and in some cases converted, from a liquid to a solid film. This terminology and the operational consequences are described in Chapter 3, but for the purposes of this market overview they may be summarised as in Table 2.

These terms are not unequivocal and may overlap to some degree; for example, powder coatings may be cured by radiation, while the latter are sometimes emulsified or dissolved in water to reduce the solids content (Fig. 2).

The largest volume of coating material used across all sectors in Europe falls into the water-borne category; however, the large decorative volume dominates this.

Table 2 Coating technology

Coating type	Description
Solvent-borne	Coating in which the volatile component is predominantly a non-aqueous solvent. Most traditional paints and lacquers lie in this category.
High solids	A solvent-borne coating, similar to above but with a higher solids content to meet solvents legislation.
Water-borne	A paint in which the pigment and binder are dispersed or dissolved in a continuous phase that consists essentially of water.
Powder coating	A solvent-free coating material in powder form, which after fusing by heat and possible subsequent curing, gives a continuous film.
Radiation curing (Radcure)	Liquid coating, often 100% solids content, which is converted into a solid by means of UV or EB radiation.

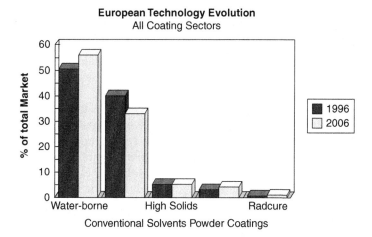

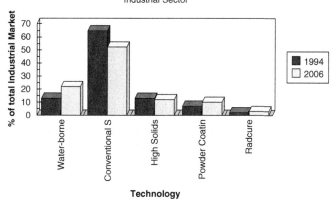

Figure 2 Technology evolution (Europe).

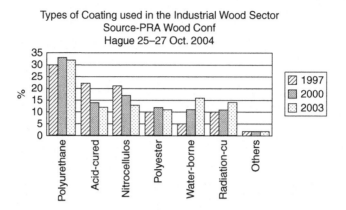

Figure 3   Industrial wood sector (Europe).

If the decorative volume is removed, then it can be seen that within the industrial sector, there is less use of water-borne technologies. Due to current and impending legislation, it can be expected that the use of conventional solvent-borne technology will continue to decline while the other 'compliant' coatings will grow.

The detailed balance of technology varies between countries and according to the market sector. Decorative wood coatings are moving strongly in the water-borne direction; however in the industrial sector for operational and other reasons, the situation is more complex and is described later. Some idea of the relative volumes is shown in Fig. 3 though it should be noted that these data conflate chemical and techno-logical terms.

2.10. Environmental legislation

The choice of technology for the wood-coatings market as illustrated above is clearly governed my many economic, commercial and opera-tional factors. In addition, there is the overarching requirement to meet current and proposed legislation. The list below shows the many legisla-tive areas that have to be taken into account (Table 3).

Legislative activity falls into three major categories:

(1) Global issues – for example, the impact of solvent on the environment including the ozone layer
(2) Local issues – for example, transport and waste regulations
(3) Personal issues – for example, the health of individuals exposed to coatings (occupational exposure)

Table 3 Areas of current legislative activity (Europe)

VOC (or 'solvents' directive)
Directive on indicative occupational exposure limit values
EU ozone depletion substances regulation
Transport and packaging regulations
CHIP 3 (chemical hazard information and proposal for supply)
    Regulations 2002
Packaging waste directive
Contaminated land
Groundwater regulations
Climate change levy
COSHH regulations (control of substances hazardous to health)
Dangerous preparations directive
Biological products regulations
Extension of the validity of ecological criteria for paints and varnishes
    eco-label
Adaptation to technical progress of the new dangerous substances
    directive
Amendment to the safety data sheets directive
Amendment to directive on plastic materials and articles to come into
    contact with food
Decorative paint directive
Coatings care/responsible care
Toys regulations
Registration, evaluation and authorisation of chemicals (REACH)

It is the first of these categories that is a major factor on technology change and is covered in 2.10.1. The impact of the solvent emissions directive has created something of a crisis for industrial wood coating and in particular furniture. Traditionally, the furniture has used large amounts of solvent in quick-drying lacquer finishes such as nitrocellulose. Some solid contents are very low and range from 20% for nitrocellulose, 32% for acid cured technology, up to 40–70% for polyurethanes and polyesters. The remainder is solvent, which will also be used for cleaning purposes. UV and water-borne technologies could in principle contain no solvent, but many water-borne technologies contain up to 10%, and UV up to 25%. Powder technology is solvent-free.

Some of the newer technologies offer specific technical advantages in terms of durability, but in many cases the challenge facing the industry is to meet the standards of appearance and operational flexibility of the older lacquers while overcoming specific problems raised by the newer technologies.

Table 4 The technology split in different European countries

Country	Technology (furniture)
France	Nitrocellulose 40–45%, polyurethane 30–40%, solvent-based UV 12–15%, Water-borne 2–5%
Italy	Polyurethanes 47%, polyesters 21%, UV 16%, Water-borne 9%, acid-catalysed ~1%
Spain	Nitrocellulose 2%, polyurethanes 40%, polyesters 28%, little or no water-borne, acid-catalysed 10%
UK	Nitrocellulose 30% (falling), polyurethanes 5%, polyester 28%, UV 5%, acid-catalysed 37%, UV 5%, water-borne 20%
Netherlands[a]	Nitrocellulose 40%, polyurethane 31%, water-borne 12%, UV 9%, acid-catalysed 7%

[a] These figures refer to clear lacquers; opaque coatings have a different balance.

Countries have reacted differently to the impact of legislation and have also different starting points. Some idea of technology differences between specific countries is shown in Table 4 (compiled from ref. [5]).

These figures are subject to the caveat already expressed that there is some overlap between the categories; the detailed implications and manufacturers preferences are discussed in Chapters 7 and 8.

2.10.1. Timetable for implementation of the solvents directive

The solvents directive [10] has an objective to reduce VOC emissions by 60–70% relative to a 1990 baseline. The options facing industry are to use compliant coatings or to destroy emissions by abatement techniques. Otherwise, production will need to sub-contracted or moved out of Europe where cheaper manufacturing costs and possibly more lenient regulations might apply. It may be noted that there are different interpretations of legislation around the world, and different interpretations of VOC. In the United States for example, solvents might be exempt if they are deemed not to be photo-catalytically active, for example, acetone. In Europe, the definition of VOC is based on a boiling point and/or vapour pressure criteria. VOCs may be expressed in terms of the grams of volatile in a litre of paint (g/l) or as the emission of a total coating process. A further complication is that water is sometimes excluded from the VOC calculation, which has the effect of increasing the g/l figure.

Detailed regulations may also depend on the size and output of the company concerned. The Europe-wide solvent threshold is 15 tonnes per year, though Germany and the UK have set the thresholds at 5 tonnes.

New installations must already comply with the latest directive but existing companies come into its scope from October 2007. However, this means that plans must already be in place to make the necessary changes. The thresholds at which the directive must be applied may differ in each country but the overall consequences are similar.

Clearly, the solvents directive (European Directive 2004/42/EC) will have different implications for decorative and industrial market sectors. For the former, there is no possibility of abatement and VOC emission limits must be applied to each product. These are usually expressed in grams per litre (g/l). Some examples of the proposed limits for the two phases of implementation are shown in Table 5. The consequences from the perspective of coating formulation are discussed later in Chapter 5.

The situation for industrial coating and in particular furniture is more complex with numerous detailed differences between countries [11]. In many cases, the approach has been to adopt a 'Solvent Management Plan', which will in principle enable the sector to meet overall emission targets. Such a plan must address fugitive and directly monitored emissions and might be addressed by a combination of abatement techniques and compliant coatings. With industrial processes, there is more interest in emissions from the perspective of the area coated (e.g., g/m^3) rather than in terms of the emissions per litre (g/l). However, technologies with an intrinsically lower volatile solvent content will clearly have an overall advantage – subject to performance and operational criteria as discussed in later chapters.

Table 5 European Directive 2004/42/EC – VOC limits for different coatings

Coating type	Max VOC limit (g/l) (ready for use)	
	Phase 1 (Jan 2007)	Phase 2 (Jan 2010)
Water-borne paints for wood trim/cladding	150	130
Solvent-borne paints for wood trim/cladding	400	300
Water-borne wood stains and lasures	150	130
Solvent-borne wood stains and lasures	500	400
Water-borne minimal build[a] wood stains	150	130
Solvent-borne minimal build[a] wood stains	700	700
Water-borne two-pack reactive coatings	140	140
Solvent-borne two-pack reactive coatings	550	500

[a] See European standard EN 927-1.

REFERENCES

[1] Akzo-Nobel's Global Coatings Report 2006 (Euromonitor).

[2] Buysens, K. (2001). Working on wood. *Polym. Paint Colour J.* **191**(4445), 12.

[3] Jotischky, H. (2007). Decorative coatings: In search of profitability and growth. *Waterborne High Solids Coat.* **28**(8), 18–28.

[4] Consolidation in the market for windows. (2004). *DIY (Europe)* **5/6**, 26–27.

[5] Jotischky, H. (2006). European furniture coatings industry at the crossroads. *In* "Coatings Yearbook 2006", pp. 159–165. Vincentz Network/International Paint & Printing Ink Council, Hannover.

[6] Jotischky, H. (2007). Beyond the compliance crossroads: The European wood coatings market on the March. *Waterborne High Solids Coat.* **29**(6), 15–22.

[7] Morrison, S. (2007). Walk this way: Floor coating technologies. *SpecialChem4Coatings* (http://www.specialchem4coatings.com/resources/articles/printarticle.aspx?id7555).

[8] Flooring: Growing happily. (Bodenbelage: Zunehmend heiter) (2006). *DIY (Ettlingen)* March (3/2006) pp 62–64.

[9] Jotischky, H. (2008). Woodcare on the go: Creative marketing, proliferating products. *Waterborne High Solids Coat.* **29**(10), 16–25.

[10] Council Directive 1999/13/EC of 11 March 1999 on the limitation of emissions of volatile organic compounds due to the use of organic solvents in certain activities and installations (amended by Directive 2004/42/EC – "Paints Directive").

[11] Secretary of State's Guidance Note for Wood Coating (defra). Progress Guidance Note 6/33(04).

CHAPTER 2

Wood and Wood-Based Substrates

Contents

Wood Coatings: Theory and Practice
DOI: 10.1016/B978-0-444-52840-7.00002-3

1. INTRODUCTION

To effectively coat wood and its derived products, it is necessary to understand the nature of the substrate. All species of wood have some features in common which may be described as 'generic', but there are also numerous detailed differences that must be taken into account during the selection and application of coatings. An analogy might be drawn with metals. Metals are conveniently defined as substance with high electrical conductivity, lustre and malleability, which readily lose electrons to form positive ions (cations). However in order to coat metal, it is necessary to understand detailed differences between (say) steel, aluminium and zinc including their production history and alloy structure.

Wood has been defined as the secondary xylem of trees and shrubs, lying beneath the bark and consisting largely of cellulose and lignin. This does not tell us much about its coating potential and it is necessary to explore other generic characteristics such as the differences between:

- Hardwoods and softwoods
- Heartwood and sapwood
- Springwood (earlywood) and summerwood (latewood)

These terms reflect biological differences, which also have chemical consequences. Although these generic terms are useful, they are only part of the story. To give an example, oak (*Quercus robur*) and balsa (*Ochroma lagopus*) are both classified as hardwoods even though they are very different in character (see Section 4). For some purposes, it is necessary to have detailed information about the species of wood as well as understanding the consequences of the generic similarities.

Wood is used both as a structural material and for decorative purposes in the form of veneers. In addition to solid wood, other materials are also used as structural elements (e.g. wood-based panels) or as covering elements (e.g. veneers, paper sheets or plastic foils). In this chapter, a short presentation of all these materials is given with special emphasis on the consequences for coating. However, some materials not requiring a coating treatment have been included to give a complete survey of this subject.

2. WOOD AND TIMBER

Wood is a complex material of plant origin and the primary constituent of trees. Timber is the name given to wood used for building constructions.

After cutting and seasoning, wood is used for many different purposes. Wood has been an important construction material since humans began building and remains in plentiful use today. In addition to building, wood has found widespread use in applications including furniture, weapons, musical instruments, domestic tools and many others.

The structure of timber can be considered at both macroscopic and microscopic levels when different features will appear. In the following sections, the composition of wood is considered starting from its chemical constituents and then looking into its morphology and structural elements. It should be stressed that there are over 30,000 different species of tree, which will vary greatly in the arrangement of the elements described below.

3. CHEMICAL COMPOSITION

The chemical substances composing wood can be broadly divided into two classes: macromolecular substances and lower molecular weight substances of both simple and complex chemical structure, collectively known as extractives.

3.1. Macromolecular substances

Wood is a composite material composed of fibres of *cellulose* (40–50%) and *hemicellulose* (20–30%) held together by a third substance called *lignin* (25–30%). Although these substances can be considered as mainly macromolecular wood components, other polymeric substances are also present in small amounts as starch and pectin derivates.

3.1.1. Cellulose

Cellulose is an organic polymer belonging to the polysaccharides family (poly = many and saccharides = sugars). The monomer is a simple sugar, namely glucose (see Fig. 1).

The structure of cellulose, presented in Fig. 2, is linear being constituted, for plant-derived cellulose, of between 7000 and 15,000 glucose units.

The glucose unit is in the β-form, depending defined by the relative position of the three hydroxyl groups in the ring. In the α-form, glucose condenses to form a starch.

Cellulose is the main constituent of the cell walls of wood fibres (the name cellulose derives from this evidence). It is a highly polar substance

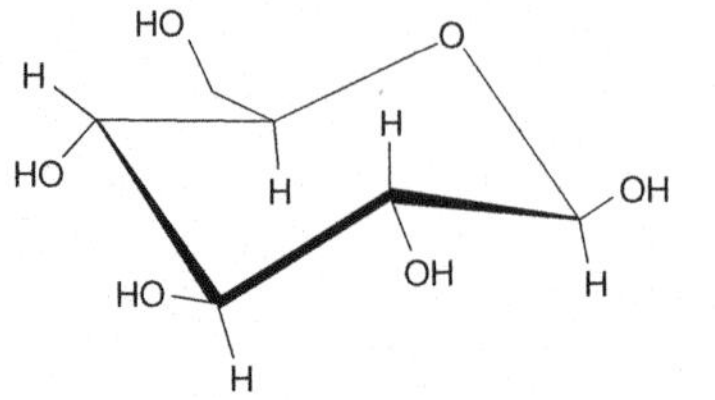

Figure 1   Glucose.

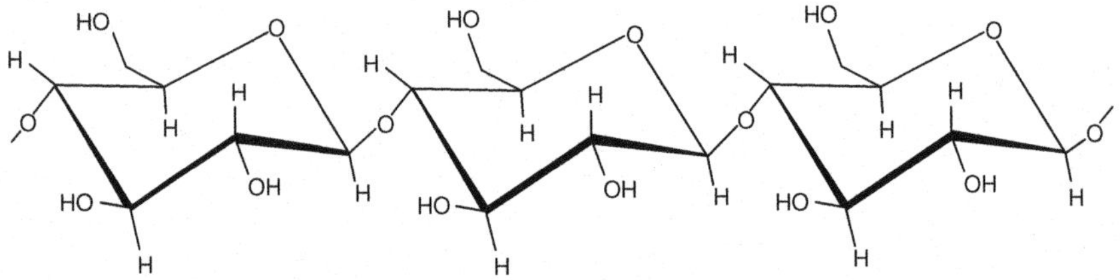

Figure 2   Cellulose (the carbon atoms in the ring vertexes have been omitted) to simplify the formula.

due to the presence of three –OH groups for every structural unit. These chemical groups give cellulose a strong affinity for water but, in consequence of the great dimensions of the polymer chains, cellulose can only absorb water without dissolving in it. Water is bonded to cellulose by means of hydrogen bonds along the polymeric chains (Fig. 3).

The interactions between water and cellulose are not so strong as the covalent bonds between two atoms of a molecule. The bonding of water to cellulose is thus a reversible process. Depending on the climatic conditions, air humidity in particular, cellulose can absorb or release water from or to the environment with important consequences for the products

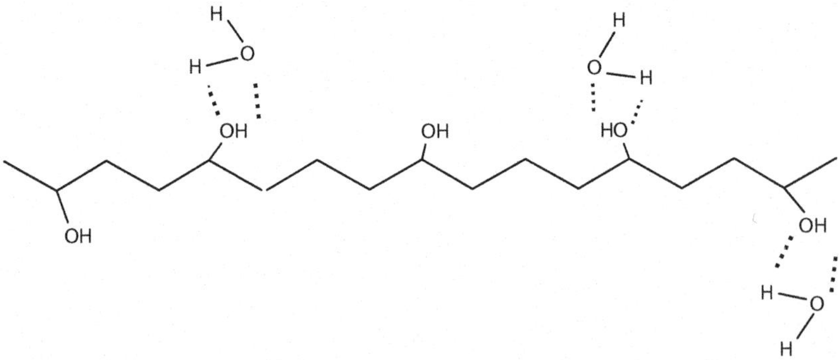

Figure 3   Interactions (hydrogen bonds) between cellulose and water.

made of wood and the role of coatings in controlling the passage of moisture into and out of coated wood.

3.1.2. Hemicellulose

In contrast with cellulose, hemicellulose is a branched carbohydrate copolymer in which the monomers are different sugars. The molecular weight of hemicellulose is significantly lower than that of cellulose. Hemicellulose acts as a cement between cellulose and lignin contributing to strength and stiffness.

3.1.3. Lignin

Lignin is an important constituent of wood representing the 25–30% of its total weight. It is a three-dimensional polymer in which the monomer can be considered the phenyl propane, with one or more methoxy groups (–OCH3) bonded to the aromatic ring (Fig. 4).

Lignin belongs to the polymer family called 'polyphenols'.

The different structural units are bonded together either by carbon–carbon or by ether C–O–C covalent bonds (Fig. 5).

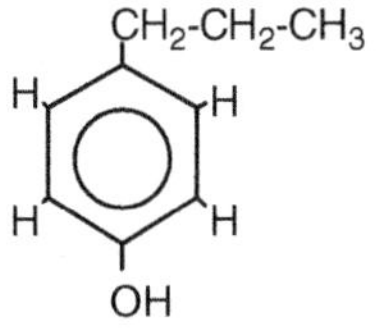

Figure 4   The phenyl propane molecule.

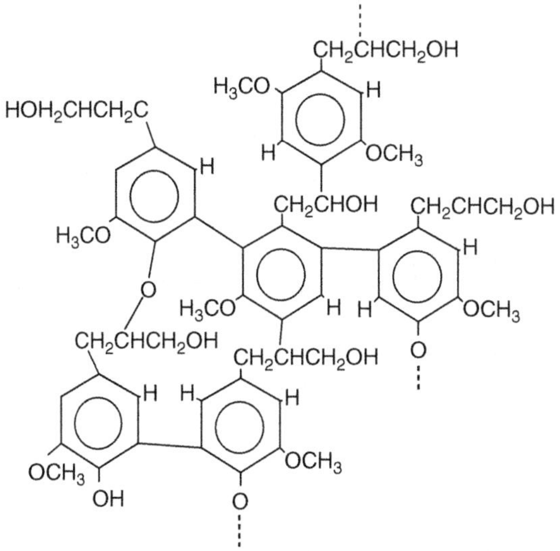

Figure 5   Hypothetical structure of lignin.

Figure 6   Examples of wood degradation.

Lignin is primarily found within the secondary cell wall and is largely responsible for the stiffness of dry wood; this is in contrast with normal cellulosic plants. The UV portion of sunlight causes photo-degradation of lignin. Degradation initiates with the formation of low molecular weight polyphenol substances which evidence is given by the darkening of the wood surface exposed to sunlight. The subsequent bleaching of such soluble compounds is caused by water in its different forms (condensation, rain, snow or frost). Wood tends then to become lighter in colour (grey) as solely cellulose remains on the surface. The final phase of wood degradation is erosion of the cellulosic structure leading to complete destruction (Fig. 6). A prime function of coatings is to protect against this process, a task that is more demanding if the coating is transparent.

3.2. Low molecular weight substances (extractives)

There are many extraneous substances present in wood (Table 1). They include both inorganic (oxides, salts) and organic compounds which vary among the different wood species in quantities that range from 5% to 30%. The organic compounds are collectively known as 'extractives' as they can be extracted from wood by using an appropriate solvent. They account for the colour and smell of wood and have important implications for coating processes. Although some extractives perform important metabolic functions, others are waste products and their presence in heartwood confers useful protective properties particularly decay resistance.

Extractives can be divided, according to their chemical composition, into three major sub-groups:

(1) Aromatic phenolic compounds
(2) Aliphatic compounds (fats and waxes)
(3) Terpenes and terpenoids

3.2.1. Aromatic phenol derivates

Phenolic compounds are second only to carbohydrates in abundance in wood and include complex polyphenolic molecules. They are principally found in heartwood and responsible for the deeper colour and decay resistance. Phenolic compounds can be sub-divided into four groups:

(1) Lignans
(2) Stilbenes
(3) Flavonoids
(4) Tannins

Lignans are stable and colourless in contrast to stilbenes which darken in the presence of light, darkening of pine is a good example caused by pinosylvin. The latter chemical can be reacted with some aromatic amines as a test for the presence of heartwood in pine.

Reactive *stilbenes* cause staining and drying inhibition of coatings on tropical hardwoods, for example, on Merbau and Iroko.

Flavonoids contain a dibenzyl propane unit and are the principal colouring material in trees, plants and flowers. They are mostly found in the heartwood of trees. Polymeric flavonoids are also known as 'condensed *tannins*' and are water extractable and a major cause of staining problems when over coated with water-borne coatings.

Another type of reactive polyphenol contains an unusual seven-carbon ring known as a *tropolone*; thujaplicins are one of the best-known examples, they can chelate with iron to give black/blue staining. Thujaplicins are the reason that Western Red Cedar is valued for its durability and fragrant odour.

3.2.2. Aliphatic derivates (fats and waxes)

The major aliphatic compounds in both softwoods and hardwoods are fatty acids, and their esters, in particular the triglycerides of saturated and unsaturated acids including linoleic, oleic and linolenic acids. Tall oil (from the Swedish word for pine) fatty acid (TOFA) is by-product from papermaking and an important component of alkyd paints.

3.2.3. Terpenes and terpenoids

Terpenes are chemically described as two isoprene units linked head to tail and found largely in softwoods. Examples include pinene, which gives pinewood its characteristic smell. The materials are volatile and 'turpentine' was once a major paint solvent and also used in minor quantities to prevent oil-based paints from skinning prematurely.

Table 1 Summary of the consequences of extractives

They are the substances main responsible for the wood colour.

Tannins can react with ferrous metals to cause blackening.

In consequence of oxidation processes caused by light and oxygen, they are the main agent responsible for the colour change of wood surfaces.

They can be responsible for some adhesion problems acting at the interface wood coat (triglycerides, waxes).

They might be responsible for aesthetical defects as whitening in the pores in consequence of optical effects or incompatibility with the coating materials.

They can be dissolved by the vehicle of certain coating materials determining, when coloured, a diffuse coloration, spots or coloured stripes called 'bleedings'. In many coating processes, special primers must be used to control staining from extractives.

As a consequence of their chemical nature, some extractives can interfere with the curing process of certain coating materials leading to drying and curing problems.

On the positive side, they can inhibit biological activity (conferring resistance to the decay fungi or insect attack). Biological resistance differs greatly between wood species and the age of the tree.

3.3. Acidity in wood

Wood is generally acidic and pH values as low as 3.0 have been found in the heartwood of oak; however, most trees have a heartwood pH of 4.5–5.5. Acidity is caused by free acetic acid, which is found in wood, and can cause corrosion of metal fittings. Metallic lead over wood, for example, in church roofs, can be badly affected and should be protected with an acid-absorbing primer.

3.4. Minerals in wood

Various inorganic elements are found in wood and participate in processes such as photosynthesis of chlorophyll. Wood may also contain extraneous deposits of crystals such as calcium oxalate and silica grains. Iroko is prone to the deposition of calcium carbonate 'stones'. Mineral deposits can cause problems during the machining of wood or coating operations.

4. MORPHOLOGY

Wood in botany is defined as the xylem tissue that forms the bulk of the stem of a woody plant. It consists of different types of cells mainly placed vertically to the three axes. Wood cells play different roles in the living tree, namely:

- Providing support to the plant structure
- Enabling sap transport from the roots to the leaves
- Storing substances, including waste products, like sugars, starches and fats
- Metabolising particular substances used by the tree for different functions

The properties of wood will differ greatly from one species to another, but it is often useful to divide wood into two broad groupings: hardwood and softwood. This is a botanical distinction; by convention, the wood from coniferous trees is known as softwood and that from broadleaved trees as hardwood. A majority of hardwoods are actually harder than softwoods but this is not always so. Balsawood, for example, is defined as hardwood despite its soft physical character. Hardwoods and softwoods show significant differences in their cell structure. Complexities exist over a wide scale of magnification.

WOOD NOMENCLATURE

A tree is conventionally defined as a perennial plant with a single woody trunk. Botanists place plants into groups that have close affinities thus creating a classification with a relationship to evolutionary history leading back to common origins.

A species is a group of similar plants which can reproduce themselves. A genus is a group of species with strong similarities but not normally interbreeding. Species are named by a two-part Latinised label, the first relates to the genus and the second to a specific term, usually with a descriptive or historical connection.

Most tree species also have everyday names, which may vary, between locations (some principle wood species are listed in Tables 2 and 3[1]).

Trees form part of a major plant division (phyla) known as the Spermatophyta, this is sub-divided into the Gymnosperms and Angiosperms. The former are more primitive and include the familiar conifers (softwoods). They have a seed which is carried unprotected on leaf-like organs which form cones. Angiospermae are flowering plants and include broadleaved trees (hardwoods).

---

[1] See Appendix for further details.

Table 2 Some principle hardwoods

Everyday name	Botanic name
Afromosia	*Pericopsis elata*
Ash (European)	*Fraxinus excelsior*
Beech (European)	*Fagus sylvatica*
Elm (English, Dutch)	*Ulmus procera, Ulmus hollandica*
Iroko	*Chlorophora excelsa*
Mahogany (African, American)	*Khaya ivorensis, Swietenia macrophylla*
Meranti	*Seraya, Lauan Shorea* spp.
Oak (European)	*Quercus robur*
Sapele	*Entandrophragma cylindricum*
Teak	*Tectona grandis*
Utile	*Entandrophragma utile*
Walnut (European)	*Juglans nigra*

Table 3 Some principle softwoods

Everyday name	Botanic name
Douglas Fir (British Columbian Pine)	*Pseudotsuga menziesii*
Hemlock	*Tsuga canadensis*
Parana Pine	*Araucaria angustifolia*
Redwood (Fir, Scots Pine, Red Deal)	*Pinus sylvestris*
Western Red Cedar	*Thuja plicata*
Whitewood (Spruce, White Deal)	*Abies alba*

4.1. Cellular structures

4.1.1. Softwoods (needlewood or coniferous wood)

Softwoods are derived from conifers, so called because of the shape of the fruits (from the Latin: *conus* = cone and *fere* = to carry). As the xylem is made up mainly of tracheids, elongated conduction and support cells whose length is around 1.4–4.4 mm (*Pinus sylvestris*), softwoods usually present a uniform appearance free from large pores.

A slice taken across the trunk of a tree shows a number of characteristics which are apparent even to the naked eye. Beneath the outer and inner bark is a thin layer of active cells (Cambium) which appear wet if exposed during the growing season. It is here that growth takes place.

In temperate climates in Spring, or when growth begins, the cells are thin-walled and designed for conduction, whereas latter in the year they

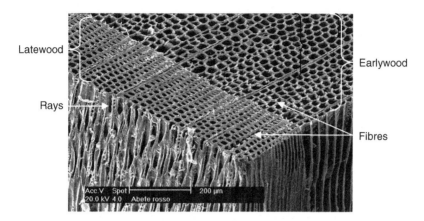

Figure 7   Wood structure – Softwood.[2]

become thicker, with emphasis on support. The combined growth forms an annual growth ring which is usually divided into two distinct areas known as springwood (or earlywood) and summerwood (or latewood) (Fig. 7). Each type has different properties and this can lead, for example, to differential movement. Division of cambial cells may be at an angle to the vertical axis creating a spiral grain formation which makes it more liable to split and twist. Growth rings give many woods a distinct 'figure' which is valued for its decorative aesthetic appearance, particular in veneers.

Resin canals, present in many types of softwood, are vertical or radial cavities surrounded by secreting epithelial cells (Table 4).

4.1.2. Hardwoods (broadleaf)

The structure of hardwoods is more complex than softwoods due to the presence of additional and more specialised cell types, short fibres (e.g. 0.6–1.3 mm in the case of Oak) for support, and vessels (pores) for

Table 4 Cell function in softwoods

Function	Cell
Support	Tracheids
Conduction	Tracheids
Storage	Parenchyma
Secretion	Epithelial cells

---

[2] SEM picture of Spruce wood (*Picea abies*). CNR IVALSA – Firenze, Italy (Simona Lazzeri).

transportation (conduction). The length of vessels can vary among species being several times longer than fibres (between 100 and 400 mm in the case of Oak). In some cases (oak, chestnut, ash), pores are quite large and distinct, in others very small. The surface porosity of certain hardwoods depends on the diameter of such cells. Hardwoods are described as 'ring porous' or 'ring diffuse' according to the distribution of the pores (Fig. 8).

The storage cells, generally grouped in small structures or tissues, are more numerous in hardwoods than in softwoods. They can be seen both in the radial and in the tangential sections looking differently according to the wood specie considered. In the case of beech, for example, they look like small and dark lens (Table 5).

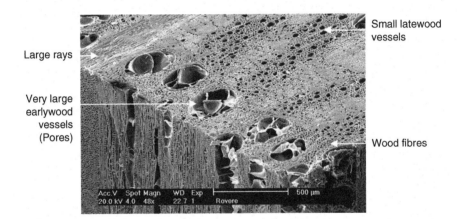

Small latewood vessels

Large rays

Very large earlywood vessels (Pores)

Wood fibres

Figure 8   Wood structure – Hardwoods.[3]

Table 5 Cell function in hardwoods

Function	Cell
Support	Fibres (and Tracheids)
Conduction	Vessels (and Tracheids)
Storage	Parenchyma
Secretion	Epithelial cells

[3] SEM picture of Oak wood (*Quercus robur*). CNR IVALSA – Firenze, Italy (Simona Lazzeri).

RESINS AND GUMS

Living trees must metabolise many substances to maintain growth, including resins and latexes. Some of these substances protect the living tree against the possible aggression of biological organisms (fungi and bacteria) or against possible mechanical damage. It is of particular interest to consider that these natural protective systems have a similar role to the treatments carried out using coating materials. Indeed, many substances are used as a component of traditional paints. In the case of pine, for example, the living tree produces a polymeric solid substance (a resin mainly derived from the abietic acid). This resin is dissolved in organic liquid substances called terpenes (the main constituent of turpentine oil, a natural solvent). When this natural mixture, reaching the damaged wood area, is exposed to air, the volatile compounds evaporate producing a protective solid shield formed by the resin. In industrial coating, resin exudation is often associated with the presence of knots. In the course of coating operations, the presence of resins and latexes may represent, in certain circumstances, a serious impediment. They can be dissolved by the coating vehicle or they may determinate a sort of barrier against an adequate contact between coating and wood. They can also migrate towards the surface through the solid coating film. Various defects may arise from the mechanisms described above, especially loss of adhesion and discolorations as these substances tend often to become yellow when exposed to light.

4.2. Heartwood and sapwood

When viewing a transverse wood section it is often noticeable that the central area is darker than the outer circumference. This is a characteristic difference between *heartwood* and *sapwood*. Sapwood is the part of the wood, the outer growth rings, which actively carry or store nutrient. Young trees produce only sapwood, but as the tree matures the cells in the centre die and become a receptacle for waste matter including tannin and gum. Once it starts to form, the growth of heartwood keeps pace with the growth of sapwood and the ratio of heartwood to sapwood gradually increases. Some heartwoods are distinctly coloured and they all show important differences to sapwood. Heartwood is less porous, shows greater dimensional stability to moisture movement and generally shows greater resistance to fungal and insect attack (Fig. 9) (An exception is an axis at the very centre of the tree, known as the pith, which is often

Figure 9   Insect attack on a piece of wood. Sapwood is almost completely destroyed whilst there is no evidence of attack on heartwood.

soft and weak). The resistance of sapwood to attack can be upgraded with preservative treatments.

4.3. Wood anisotropy

As a consequence of the morphology described above, wood is anisotropic, that is, to say its properties and appearance vary according to the spatial orientation. As wood can be cut in any direction, the result will be surfaces with different appearance and properties. The principal directions of cut will be transverse radial or tangential (see Fig. 10).

4.4. The cutting of wood

Cutting logs into sections before seasoning is known as 'conversion' whereas subsequent shaping is generally called manufacture. During cutting considerable dust is produced which may present health and explosion hazards.

Methods of cutting include sawing, peeling and slicing. Subsequent finishing operations include planing and sanding.

The final finish achieved during manufacture can play an important role in the eventual weathering of surface coatings. Sawn surfaces show

Figure 10   Wood – the three directions of cut (X = transverse, R = radial and T = tangential).

more absorption than planed ones; this can be advantageous in the cases of penetrating stains. Sawn surfaces also seem able to redistribute some surface stresses which is advantageous to coatings. When wood is planed the surface may become compressed, particularly if the knives are blunt. Later release of this compressive stress is detrimental to the performance of paints.

The two main types of cut are usually known as 'plain sawn' and 'quarter sawn' defined as follows:

- *Plain-sawn timber* (flat grain timber, flat-sawn timber, slash-sawn timber). Timber converted so that the growth rings meet the face in any part at an angle of less than 45°.
- *Quarter-sawn timber* (rift-sawn timber, edge grain timber, vertical grain timber, comb grain timber). Timber converted so that the growth rings meet the face at an angle of not less than 45°.

Plain-sawn softwoods usually have a more decorative appearance but quarter-sawn boards shrink less in their width (see moisture movement) and are therefore less liable to cup or twist. In flooring quarter-sawn boards give more even wear. The grain raising of quarter-sawn timber is usually finer than plain sawn. Weathering of coatings for exterior use will also be different and for this reason the European weathering test method (EN 927-3) gives a detailed specification for a simple panel test.

4.5. Wood and water (moisture content)

Water is vital to the living tree, and the weight of water is usually greater than that of the dry wood. After felling and seasoning, the tree loses most of this water, but because wood is hygroscopic it retains a moderate amount which continues to have an important bearing on many of the wood's properties, including strength. Water occurs in living wood in three conditions:

(1) In the cell walls
(2) In the protoplasmic material of the cells
(3) As free water in the cell cavities and spaces

In common with other hygroscopic materials, cellulose absorbs different amounts of water according to temperature and humidity and, in so doing, changes its dimensions. This is significant and must be allowed for in the design of structural units.

The moisture content of wood is conventionally expressed as a percentage based on the weight of dry wood. When newly felled, timber can contain up to 200% moisture contained in cell cavities and walls. As wood is dried or seasoned water is lost from the cell cavities, *with no corresponding change in volume*. Eventually the point is reached where cavities contain no liquid water and remaining water is held in the cell walls; this is known as the *fibre saturation point* which occurs around 30% moisture content; further loss causes dimensional change. Wood will establish an equilibrium moisture content dependant on the temperature and humidity of the surrounding air.

It is customary to describe the changes in dimension which occur in drying timber from the green state as *shrinkage*, while changes which occur in seasoned wood in response to seasonal or daily fluctuations are known as *movement*. Typical shrinkage values are 0.1% longitudinally and up to 10% tangentially; radial movement is around half that of the tangential. By convention, movement is reported as a percentage change occurring between 90% and 60% RH at 25 °C. Tangential movement can be up to 5% but differs considerably between species (see Table 6).

Timber in service is subjected to a constantly fluctuating environment both indoors and outdoors. Factors include changes in humidity both internally and externally, and the effects of rainwater and condensation. This will lead to movement, and when considering the applicability of specific coating types it is useful to divide wooden components into those such as joinery where dimensional stability requires control, and those like fencing, where movement is less critical. The rate at which movement occurs depends on the water permeability of the coating and this is an important factor to be considered when designing a wood coating system. Prior to coating, the moisture content of timber should ideally be allowed

Table 6 Percentage movement from 90% to 60% relative humidity

Wood species (density[a])	Moisture content change (%)	Tangential movement (%)	Radial movement (%)	Movement classification
Teak (650 kg/m^3)	5.0	1.2	0.7	Small
Pine (570 kg/m^3)	7.5	2.1	0.9	Medium
Oak (760 kg/m^3)	8.0	2.5	1.5	Medium
Beech (750 kg/m^3)	8.0	3.2	1.7	Large

[a] Density, in brackets, is reported as medium value for wood with 12% moisture content.

to reach an equilibrium compatible with those conditions under which it is to be used. A number of standards and technical publications describe the required moisture content in detail (see [1–2]).

4.6. Seasoning of wood

The purpose of seasoning is to render timber stable at a moisture content appropriate to the intended end use. It is therefore necessary to bring timber in a controlled manner from the high moisture contents of newly felled wood, to the average equilibrium conditions that will be required in service. Seasoning requires balancing the rate that water is lost from the surface with internal movement of water from the interior. If this is not under control, various seasoning defects can occur such as 'warping', 'twisting' and cupping. Internal rupture can lead to splitting and other internal defects. Timber was traditionally seasoned by air drying and this is still practised in some countries; however, the majority is now kiln dried.

4.7. Dimensional movement of wood

As described in the previous sections, cellulose is the main constituent of the wood cell walls. In absence of water, the polymeric chains are closed and entangled each other in consequence of the polar interactions deriving from the presence of several hydroxyl groups constituting the glucose structural units.

The presence of water in the form of atmospheric humidity causes interactive competition since water is itself a polar substance. Water is able to penetrate through the cellulose chains binding to them and reducing the interactions among the polymer chains. The effect on wood structure is swelling of the cell walls. This competition between cellulose and water can be considered as a dynamic reversible process in which the equilibrium value is constantly changed according to climatic variations. As air humidity increases, an increase of the interaction between water and cellulose occurs. Conversely, a dry climate favours restoration of interactions among cellulose chains, and the consequent release of water by wood. The macroscopic effects of such a complex mechanism are continual dimensional variations of wood. An expansion of wood elements will occur as the humidity changes from dry to humid with a contraction in the opposite situation (Fig. 11).

This property is fundamental for wood components as it can cause either functional defects or aesthetical drawbacks (e.g. warping and splitting).

One of the fundamental roles of the coating processes is often to control moisture movement in response to environmental change. However, it is important to stress that coatings cannot prevent movement, only the rate at which equilibrium is attained. It is therefore incumbent upon coating systems to have adequate mechanical properties, so they are not themselves damaged by moisture movement of the substrate. As noted later this requires the coating to have a high extensibility, relatively low modulus of elasticity and good adhesive performance. For external weathering, it is also important that the properties of the coating do not

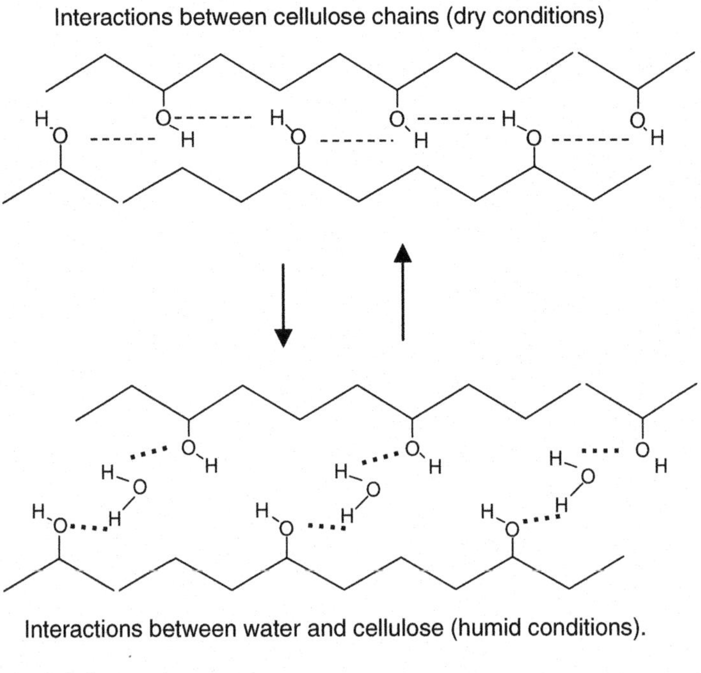

Interactions between cellulose chains (dry conditions)

Interactions between water and cellulose (humid conditions).

Figure 11 Cellulose and water.

excessively change in response to exposure as this will lead to defects such as cracking and flaking which are exacerbated by wood movement.

Taking into account the above considerations, the following points may be emphasised:

- During all the working process of wood products, the humidity content should be as close as possible to that determined by the environmental conditions of its final use. By following this simple rule, it is possible to minimise the dimensional variations that are harmful for the product functionality or appearance.
- The wood-drying processes and the following conditioning phases (transport, storage) are determining factors for all subsequent production processes and thus the final result.
- As a consequence of its anisotropic cellular structure, the dimensional variations of wood are different along the three spatial dimensions and can be described as:

  (1) Negligible along the axial direction
  (2) Significant in the radial direction
  (3) Considerable in the tangential direction

- Dimensional variations are different between wood species. As a consequence of their structure, these variations can be considerable. Denser woods take up more water due to the higher wood mass. A few examples are given in Table 6. The sum of tangential and radial movement may be used to categorise overall movement as small, medium or large.
- Coating films are not able to completely prevent wood from humidity exchanges with environment. The function of coating films is that of slowing down such process, modulating the effects of climatic variations. Permeability of coating systems towards water in its liquid or vapour states is function of the coating composition and film thickness. These properties could also vary in the course of time depending on weathering processes affecting organic polymers.
- Because coating films are not able to totally prevent moisture ingress and the consequent movement, adequate dimensional tolerances should be allowed in the designing and production of wood products to reduce the onset of defects on the final products. Tolerances must be stricter for applications such as window joinery, than cladding or fencing. EN 927-1 defines end-use categories that reflect this.

4.8. The density of wood

The actual density of wood substance is about 1500 kg/m^3; however as the result of cells and pores, etc., wood density ranges across different species from around 160 to 1230 kg/m^3 (Table 7). It will also vary within

Table 7 Wood density

Wood species	Density of anhy-drous wood (kg/m^3)	Density of wood at 12% moisture content (kg/m^3)
Teak	595	650
Pine	490	570
Oak	650	760
Beech	680	750
Balsa	160	176
Lignum vitae	1210	1230

species and individual specimens. Density is the single most important property controlling mechanical properties, especially strength. The density of wood is clearly a function of moisture content. Since the latter will change both mass and volume, they must be determined at the same moisture content. Generally, mass and volume are determined at zero moisture content (dried at 105 °C). Data are usually given for a moisture content of 12% which for many species is the equilibrium value at 65% RH. Tables of density for different humidity values are available (see [2]). An approximate rule is that density increases by 0.5% for a 1% moisture content rise up to 30%. Thereafter, the rate of increase is more rapid.

5. BIODEGRADATION OF WOOD

Wood is vulnerable to attack by bacteria, fungi, insects and marine borers. Bacteria play little obvious part in the decomposition of wood though their presence is sometimes a precondition for more serious attack. Wood that has been stored in water can become very permeable as a result of bacterial attack and this may affect the uptake of water and preservatives with consequent influence on paint properties. Fungal attack is more serious and can cause either disfigurement or structural damage, according to the species. Softwoods are particularly prone to a blue staining (also known as 'sap stain') which is caused by several types of fungi, including *Aureobasidium pullulans*. Further staining, known as 'Blue Stain in Service' can be caused by infection arising from colonisation of surfaces during or after manufacture, and is not necessarily the final stage of growth of blue stain fungi originally present in the timber. In addition to the disfiguring surface moulds and blue stain fungi, there are a substantial number of fungal species that cause serious structural damage by a process, which in its advanced stages, is known as rot or decay. Besides this direct effect, fungi can alter the surface porosity causing possible anomalous absorption of the coating materials applied to the surface.

5.1. Wood as a nutrient

5.1.1. Decay and fungal attack

Being mainly constituted by polysaccharides and other organic substances like starches and sugars, wood represents a nutrient source primarily for those living organisms able to obtain the glucose by enzymic breakdown of the cellulose chains. Water itself is also an important substance utilised for all the biological processes of the living beings and wood which is kept dry is largely immune from biological attack. A moisture content below 20% is usually considered as the safe limit for the risk of decay to be negligible. Above this limit, as would be the case for wood in ground contact some form of *preservative treatment* must be used. Soft rots including fungi from Ascomycotina and Deuteromycotina groups are limited in their growth by the availability of nitrogenous nutrient, but degraded surface layers can have an effect on coating performance. More serious in their effect on structural properties are the fungi that destroy wood. Most are from the Basidiomycetes group. They attack both lignin and cellulose and are less restricted by nitrogen availability. This group includes 'wet' rot (*Coniophora puteana*) and 'dry' rot (*Serpula lacrymans*).

5.1.2. Insect attack

Some insects called xylophagous (from the combination of the two Greek words *xylon* = wood and *phagos* = to eat) can utilise wood as nutrition source. The effects on wood of such aggressions depend on the specific nutrients that the various organisms use. The most common form of insect attack is by wood-boring beetles. They can be grouped into three major damage categories:

(1) *Category A*. These can cause serious structural damage, for example, death watch beetle, house longhorn beetle. Insecticidal treatment is usually needed to eradicate.
(2) *Category B*. Here the insects only feed on damp rotted wood and the remedial measures for controlling decay will prevent further infestation. Examples include wood-boring weevils and stag beetles.
(3) *Category C*. Only attack green or partially dry timber, or in some cases use timber as a refuge rather than food source. Insecticidal treatment is not usually necessary, for example, sawfly, wood wasp, forest longhorn beetle.

6. MODIFIED WOOD

For over a century, different wood treatments have been investigated in order to modify the natural composition of wood to produce a 'modified' material more resistant especially to the degradation factors affecting its

Table 8 Wood modification processes (Europe)

Type of treatment	References
Thermal woods	Retification (heat + inert gas) – France
	Thermowood (heat + steam) – Finland (VTT)
	Thermal oil (Germany)
Hydrothermal	Plato wood (Netherlands)
Acetylation	Microwave process – Sweden
	Uncatalysed acetic anhydride – Netherlands
Furfurylation	Furfuryl alcohol – Sweden, Norway
	Wood Polymer Technologies – VisorWood, Kebony, Kebony Dark
Ammonia (+heat)	Korespol s.r.o. – Slovakia
Bio-treatments	Fungal enzymes (Lignocell) – Germany
	Enzimes (KVL) – Denmark

structure during outdoor exposures. The value of improved properties must be balanced against extra processing and chemical costs. The research on such treatments can be summarised into thermal, chemical and surface (including biological) categories. Sometimes, also wood impregnation is classified as a 'wood modification' process although the processes of filling the cell walls with monomers, resins or oils do not truly involve alterations of the chemical composition of the wood components (Table 8).

6.1. Thermal treatments

There are several thermal processes which require heating wood to high temperatures up to 200 °C. Industrial plants use various treatments which vary in terms of temperature, duration of treatment, possible use of special gases or vapours, or immersion in certain liquids (e.g. oils). Heat causes different chemical modifications, which are not fully understood, of the wood constituents. Some of the possible reactions taking place relate to the hydroxyl groups of the cellulose chains. The degradation of some constituent (e.g. hemicellulose) could in principle produce acid species reacting then with the OH groups of cellulose producing ester groups. This process determines at the end, a reduced affinity of wood towards water. The consequence is a tangible benefit in terms of the reduction of wood swelling being also reflected in a better biological durability. On the negative side, such thermal processes can reduce other properties such as the mechanical strength.

$$\text{Wood-OH} + \text{CH}_3\text{COOH} \rightarrow \text{Wood-OCOCH}_3 + \text{H}_2\text{O}$$

Figure 12   Schematic representation of the reaction between wood (cellulose) and acetic acid producing 'acetylated wood'.

6.2. Chemical treatments

In this case, the cellulose hydroxyl groups are etherified, by the reaction with alcohols (e.g. furfurilic) or esterified by the reaction with acids or anhydrides (e.g. acetic acid or acetic anhydride).

These chemical treatments are able to reduce the sensitivity of wood towards water (Fig. 12).

6.3. Surface treatments

Surface modification of wood is used to directly improve certain properties e.g. stability by grafting UV adsorbers on its surface. Other treatments have been developed to modify its surface energy in order to make wood hydrophobic (e.g. chemical modification with silicones) or to improve its compatibility with coatings (plasma, corona discharge).

Biological modification of wood is essentially carried out by the use of enzymes.

Phenol oxidases, peroxidase and laccase are capable of producing an oxidising surface acting directly on lignin. The biological modification produces 'activated' surfaces more prone to self-bonding of wood fibres (e.g. production of wood boards).

7. WOOD APPEARANCE

7.1. Colour

The colour of wood surfaces depends mainly on the wood species. There are many different shades from the lightest whitish spruce or maple, to the darkest mahogany or ebony. In addition, there may be very significant differences between the heartwood and sapwood of a particular species (Fig. 13).

Some woods can have an intensive coloration which is valued for decorative applications including veneers. In addition, the colour of a wood surface can be modified by treatments mainly based on the application of chemicals, stains and dyes. The scope of such treatments is usually that of conferring particular colours to reproduce the appearance of certain valuable wood species.

Figure 13 Heartwood and sapwood contrast in Yew (*Taxus baccata*).

The coloured compounds in most woods are not particularly stable and will change according to circumstances:

- *Light*. Most woods change colour on exposure to light, some darken and some bleach. Teak, Afromosia and Iroko change colour in sunlight in a way that is influenced by the direction of cut.
- *Kiln changes*. Kiln drying (seasoning) and other thermal treatments (e.g. the use of vapour in certain plywood-bending processes) can cause discoloration; spacing 'stickers' may thus leave marks. Iroko is particularly prone to this problem.
- *Enzymes*. Enzyme staining (e.g. from fungal attack) may occur below the surface of the wood and become apparent on working.
- *Metal*. Iron reacts with tannin to give a blue–black stain. Oak, tropical hardwoods and Douglas Fir are among those affected.
- *Alkali*. Alkaline solutions (e.g. from cement) can cause intense colours to develop in some woods. Ammonia will stain those woods containing tannin, this can give problems with gluing and means that certain water-borne coatings will give a different colour to their solvent-borne equivalents.
- *Acid*. Acid causes pink coloration in some woods and therefore affected by acid-catalysed coatings or acid adhesives (e.g. PVA) used for the gluing of veneers.

One of the pleasing features of timber is the attractive appearance of the surface which is revealed by cutting and this, in turn, has created the demand for transparent and penetrating coatings. The decorative features of timber can conveniently be discussed under the following headings:

- *Texture.* This reflects the distribution and size of cells and other elements. Thus boxwood with its narrow growth rings is described as fine or close texture, whereas ash with very porous springwood is described as coarse.
- *Figure.* Ornamental markings on the cut surface formed by structural features such as grain, growth rings, rays and knots.
- *Grain.* This arises from the patterns of alignment taken up by more or less vertically oriented cells. Many arrangements are possible leading to terms like 'cross-grain' and 'diagonal' grain. When the alignment of cells from the vertical deviates in a constant direction, it produces a spiral orientation. Certain hardwoods such as mahogany have an alternating spiral helix which is said to be interlocked. The radial faces of such timbers are characterised by alternating light and dark longitudinal bands. The helix may also alternate in the longitudinal plane producing a wavy grain on tangential faces.
- *Rays.* Both hardwoods and softwoods may contain radially oriented cells which are specialised for communication from the bark to inner portions of the tree. Rays are especially wide in the case of Oak, giving a characteristic silver ribbon-like grain in quarter-sawn wood.

These characteristics are fundamental to the surface appearance of wood but there are also some potential problems. The use of products containing solvent and especially water can cause a differential absorption and therefore a different swelling of wood, especially when the differentiation between rings is particularly pronounced. The vessels and the resin channels cause surface porosity that might produce undesirable optical effects during coating operations if the coating material is not able to penetrate into the pores. However on the positive side, porosity increases preservative penetration into wood. The deeper the penetration, the more effective is the treatment against decay.

8. UTILISATION OF WOOD (TIMBER) IN CONSTRUCTION AND FURNITURE

Wood is used in many forms but a useful distinction may be drawn between solid wood (which may be joined in various ways) and composite products where reconstituted wood is combined with adhesives and other materials to produce panels and moulded products.

8.1. Solid wood

The use of solid wood requires selecting a wood species that meets an appropriate specification. There are many European and National standards covering different industries. The use of solid wood will often require particular attention to joining together different parts by means of adhesives and jointing techniques. Timber finds widespread use in construction due to a favourable strength to weight ratio and a high flexural rigidity. Further improvements can be made with laminated beams, which also have the advantage that they can be tailored to specific dimensions.

8.1.1. Timber grading

Timber was traditionally graded by an appearance system based on visual defects; however in the case of softwoods, this has been increasingly superseded by mechanical stress grading. Stress grading relies on a correlated relationship between modulus of rupture and modulus of elasticity. Stress grading of hardwoods is also possible but less well established.

8.2. Wood-based panel products

Products derived from wood contain wood in a number of different physical forms combined with glues, resins and other additives. Such products are usually produced as panels; some are also suitable for mouldings. The coating characteristics of these derived products may be significantly different from the parent wood and many of them are not suitable for exterior use. A big advantage of wood-based panel products is that they can utilise a wide range of woody materials including lower grade and waste products. However, the major source of raw material is plantation grown softwood. The main categories of wood-based panels are:

- Solid wood panels (including blockboard and laminboard)
- Plywoods
- Particleboards (chipboards)
- Oriented strand boards
- Fibreboards

8.2.1. Solid wood panels

According to the definition given by the European standard EN 12775 (Solid wood panels – Classification and Terminology), solid wood panels consist of pieces of timber of longish form glued edge to edge and, if multi-layer, face to face. The 'pieces of timber' forming such panels are defined as boards, lamellas, blanks, strips or planks depending on their dimensions (Table 9).

There are many types of wood-based panels belonging to this definition. The term 'blockboards' is commonly used for panel derived by assembling wood strips to form a flat surface (e.g. typical use: window

Table 9 Solid wood panels

Pieces of timber	Dimensions
Lamellas	3 mm ≤ thickness ≤ 10 mm and width ≥ 25 mm
Boards	10 mm ≤ thickness < 40 mm and width > 80 mm
Strips	Thickness < 40 mm and width < 80 mm
Blanks	Thickness > 40 mm and width ≤ 80 mm
Planks	Thickness ≥ 40 mm and width > 80 mm

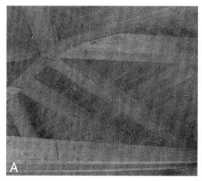

Figure 14 Example of solid wood panels: (A) roof built with laminboards and (B) furniture elements produced using blockboards.

frames). Laminboards represent also an important class of solid wood panels much used in building construction (carpentry) (Fig. 14).

8.2.2. Plywood
The term 'plywood' derives from the Latin *applicare* = to attach, to apply.

Plywood was used also in ancient times; there is evidence of the use of such materials in Egypt around 3500 BC. The term includes different types of panel which common property is the evidence of being constituted by different layers. European Standard EN 313 classifies plywood panels according to the construction and use as an arrangement of panels which consists of an assembly of plys bonded together. The direction of the grain in alternate plys is usually at right angles. The outer and inner plys are generally placed symmetrically on both sides of a central ply or core. Plywood is sub-divided into veneer plywood and core plywood, the latter may be sub-divided (see ISO 2426) into:

• Wood core plywood
• Cellular plywood
• Composite plywood

The simple plywoods are those in which the layers are made of thin sheets of wood, called plies, which are stacked together with the direction of each ply's grain differing from its neighbours by 90° (cross-bonding). The plies are bonded under heat and pressure with thermosetting adhesives, usually urea–formaldehyde for interior use or phenol–formaldehyde for exterior uses. The performance is very dependent on the quality of the glue. A common reason for using plywood instead of plain wood is because these panels re more stable and because they are less prone to dimensional defects (shrink, twist or warp). 'Core plywood' describes panels in which the core, between the outer layers, is made of other wood-based panels or other materials.

8.2.3. Particleboards

Particleboards are widely used as flat panels for the construction of furniture elements. The production of particleboards started during the early 1940s in Germany as a possible reutilisation of wood waste. They are obtained under pressure and heat from particles of wood with the addition of polymeric adhesive. The process can be obtained by:

- Flat pressure
- Drum pressure
- Extrusion

As wood represents the main constituent (over the 90%) of this material, some of the fundamental properties of particleboards can be considered similar to those of wood but others, more specifically related to the structure, are considerably different. Although chipboard can be coated directly, the result is not very satisfactory. More frequently, they are covered with various materials, wood veneers, paper sheets, plastic foils that can be followed by the application of one or more coats. European standards EN 309 and EN 312 (divided in different parts) represent a useful source of information about the classification and properties of such panels.

8.2.4. Fibreboards

Fibreboards are obtained by hot pressure of wood fibres. The cohesion among the fibres can be obtained by their own intrinsic cohesion or by the addition of an adhesive (usually an urea–formaldehyde glue). The structure of fibreboards is more uniform and compact allowing to be worked in a way more similar to solid wood. The use of fibreboards is particularly recommended when making rounded parts with different profiles or when the substrate is moulded. As these panels are more versatile, there are several possible applications. From the point of view of finishing, they are preferred to particleboard for direct use of coating materials (mainly opaque coatings in this case). They can also be covered with veneers and other materials described above, before being subject to a final coating process. The definition and classification of fibreboards are given by EN 316. This European standard classifies those panels according to different criteria. The first is related to the production process:

- Panels produced by wet process (presenting a fibre humidity over the 20% during their formation).
- Panels produced by a dry process (presenting a humidity below the 20% during their formation and a density ≥ 450 kg/m^3). These panels are called MDF (medium-density fibreboard) and represent the most useful class of fibreboards for the furniture sector.
- Fibreboards are further divided according their density and other supplementary properties (final use and use conditions).

The properties of fibreboards are described by another European standard, the EN 622, also divided in different parts. Fibreboards may be combined in layers to form 'sandwich' structures with a more cost-effective balance of surface and strength properties. As the use of MDF in furniture increases, there has been a growing need for tighter specifications to cover service life and interactions with the coating during application and curing. A list of properties related with furniture coating processes is presented in Table 10.

Table 10 MDF properties which should be considered for coating processes

Strength properties (flexural strength + modulus)	Requirement depends on end use (e.g. flat work tops vs cabinet doors)
Resistance to tensile strength perpendicular to surface	Sufficient cohesion of the outer layers can prevent defects due to tensions induced by coating drying (volume contraction)
Swelling in water	Important for kitchen and bathrooms applications
Toluene and water test (surface absorption)	A measure of coating absorption
Panel hardness	Important for the surface properties of the finished product (e.g. resistance to impact)
Density profile	Density shall be balanced. Important property also for possible moulding processes
Residual stress	Important for the planarity of the coated panel
Board conductivity	Important property related to electrostatic spraying application (powder coating)
Degassing temperature	Important property for powder coating application

8.2.5. Multi-laminar wood

Multi-laminar wood is produced by gluing together many thin wood layers previously coloured, according to the desired final result, to produce a multi-layer woodblock. By cutting this block at different angles, new wood layers having different surface patterns are exposed. The process can be repeated combining foils derived from previous processes. The complexity of colour and patterns can also be designed with the aid of computer aided (CAD) techniques. Two main features are:

(1) The final multi-laminar product can reproduce the grain patterns and figure of a large variety of wood species.
(2) It may produce wood-based material with effects and colourings rare or non-existent in nature.

9. COVERING MATERIALS

9.1. Decorative veneers

The technique of veneering was known 2000 years ago to both the Egyptians and ancient Greeks; it was again revived in the Renaissance. Veneers remained a luxury item until the modern age. Veneers are essentially thin slices of wood the appearance of which will depend on the way in which a log is offered to the cutting tool. Veneers are cut by peeling (on a lathe) or by flat cutting with a piston operated knife. Decorative veneers are primarily produced by flat cutting whereas the construction veneers used for plywood are invariably peeled. The patterning or figuring of a veneer will clearly depend on the species and the direction of cut. Sometimes, outgrowths can be cut to produce highly desirable effects as in burr veneers. After cutting the slices are re-stacked to avoid major grain variations between sheets. Furniture manufacturers will purchase by the stack, or from a closely aligned set. Veneers also can be obtained from multi-laminar wood as described above. Veneers can be joined together to produce wider surfaces. The joining process is usually carried out by means of thin plastic wires (e.g. nylon). They are partially melted and applied by zigzagging across the joining line of two veneer foils. The thickness of veneers usually varies from 0.3 mm up to 5 mm then applied by bonding to the surface of wood-based panels by means of urethane, ureic or vinyl adhesives (also mixed together). The adhesive partially fills the pores of the veneer and presents a surface with excellent finishing properties. Veneered panels are then subject to coating processes using mainly transparent coating materials.

9.2. Impregnated papers

Impregnated papers can be produced in different colours and patterns (including wood patterns), and due to their relative low cost and versatility are often used as covering materials for wood-based panels. There are two main classes of papers, usually called the 'melamine' and the 'finish'.

9.2.1. Melamine impregnated papers

The structure of melamine impregnated paper is illustrated by Fig. 15. The paper is impregnated with a thermosetting melamine or ureic resin to close all the paper pores. Partially cured melamine resin is then spread on both faces of the paper achieving bonding and curing in the same process, when applied to a wood-based panel, using heat and pressure. Two European standards, EN 14322 and EN 14323, describe the test methods and the performance levels to be reached by particleboards or MDF faced with melamine papers. These panels are usually 'ready to use' without a further coating process.

9.2.2. Finish impregnated papers

The structure of the finish impregnated paper is illustrated by Fig. 16. In this case, the paper is impregnated with thermosetting resins (e.g. ureic) usually in combination with an acrylic resin conferring 'flexibility' to the paper. The outer layer can consist of an acrylic resin suitable for further coating, with different coating materials. Alternatively, the outer layer should be characterised by high-performance levels without further coating. The finished impregnated paper shall be applied on flat or rounded elements by using an appropriate adhesive.

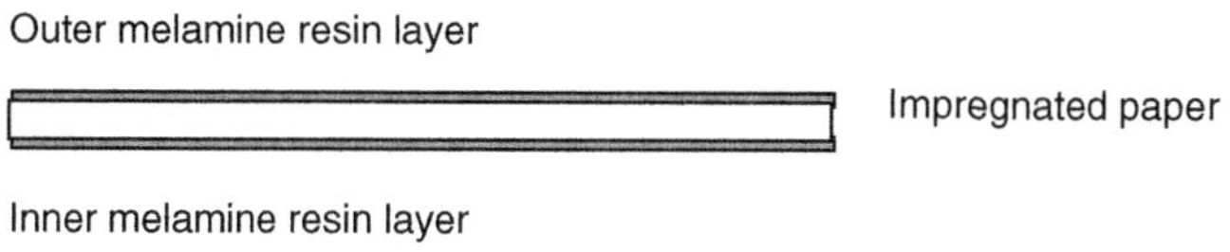

Outer melamine resin layer

Impregnated paper

Inner melamine resin layer

Figure 15   General structure of melamine papers.

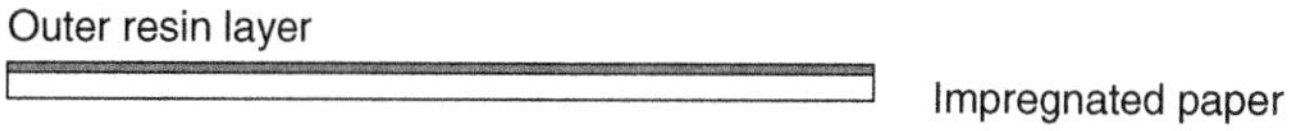

Outer resin layer

Impregnated paper

Figure 16   General structure of finish papers.

9.3. Plastic sheets

Plastic sheet can be directly applied onto panels surface gluing. Polyvinyl chloride (PVC) is the most widely used plastic material. It is a thermoplastic polymer obtained by a poly-addition reaction of vinyl chloride monomer.
    The success of PVC derives from the following properties:

- Availability in many colours
- Relatively low cost
- Thermo-formable
- Resistant to exterior weathering (unplasticised PVC is also used as structural material for the construction of window frames)

PVC foils are often covered with a thin layer of a coating film. Wood-based panels covered with PVC foils are also used as a stable and uniform base material to be coated with pigmented coating systems. Other polymeric materials can be obtained in thin foils and used as covering materials for wood-based panels. Polypropylene, a polyolefin obtained by polymerisation of propylene, also shows good performance. Other plastic materials such as ABS (a copolymer acryl butadiene styrene) or polystyrene are generally used in narrows laminas to cover the panel's edges of furniture elements.

9.4. Laminates

A laminate is a material constructed by uniting two or more layers of paper together. The process of lamination usually refers to a sandwiching process involving heat and pressure. Laminates, usually not subject to further coating processes, are classified according to the process used for their production.
    Two major categories are defined by the process of manufacture.

9.4.1. High-pressure decorative laminate (HPL)
HPL classifies laminates consisting of core layer of papers, usually impregnated with phenol resins, and the outer layer of melamine or other resins. The high-pressure process requires the simultaneous application of heat ($t > 120\ ^\circ C$) and pressure (≥ 7 MPa) to provide flowing and subsequent curing of the thermosetting resins to obtain a homogeneous non-porous material with increased density (≥ 1.35 g/cm^3). Some laminates are produced with a further protective surface overlay. HPL thickness usually varies between 0.5 and 1.0 mm with different categories of appearance (roughness, colour and gloss). EN standard 438 defines and classifies such materials.

9.4.2. Continuously pressed laminates (CPL)
CPL denotes materials consisting of paper core layers impregnated with phenol or amino plastic resins and a paper surface layer impregnated with amino plastic resin (usually melamine). Layers are continuously

bonded together by means of a laminating process utilising heat and pressure. The outer layer on one side has decorative colours or design. They are produced in rolls with thicknesses around 0.2 mm, and weight around 200 g/m^2.

APPENDIX: SOME IMPORTANT WOOD SPECIES (REFS 7-8)

Afrormosia is a yellow–brown hardwood very similar in appearance to teak which is mainly grown in Ghana and the Ivory Coast. It is used for furniture interior and exterior joinery and cladding. It is a very durable wood; however, it does tend to darken on exposure and in damp conditions it reacts with iron and steel fixings, etc., which causes the wood to turn black. It is particularly suitable for staining with wood stains.

Afzelia is a reddish-brown hardwood which is imported from West Africa. It is a very durable wood used for interior and exterior joinery and also cladding. It exudes a yellow dye in damp conditions which can be a major problem in terms of staining. It can be difficult to obtain a uniform stain appearance on those areas which show scattered groups of pores containing the yellow/white deposit. Afzelia is very resistant to decay and chemical attack.

Agba is a yellowish-brown hardwood from West Africa. It is a particularly stable wood which is very resistant to decay and chemical attack. It is used for cladding, trim and joinery work in the building industry. However, it can be a very troublesome wood due to gum exudation and occasional large resin pockets. However, it can give good results when stained if it is properly filled with a suitable coloured wood stopper.

Ash is a hardwood which is white to light brown in colour and widely grown in Europe and South West Asia. When first cut the timber turns pink in colour. It has a straight grain with conspicuous growth rings. Ash has good steam-bending properties and is often used for interior joinery. It is particularly suitable for staining and varnishing.

Beech is a hardwood somewhat variable in colour ranging from a whitish colour to pale brown and pinkish red when steamed (e.g. for use in sports equipment, etc.). It is grown in Europe and Western Asia and has straight grain with conspicuous growth rings. It is used for interior joinery flooring and plywood as well as furniture and sports equipment. Beech is not a very durable wood and so not suitable for exterior use. It is particularly suitable for staining with wood stains.

Birch is a hardwood from Europe and North America. It ranges in colour from white to reddish brown. It is not a very durable wood and has a plain appearance. Birch has excellent bending properties and is used for interior plywood and flooring as well as furniture making. It is particularly suitable for staining with wood stains.

Cedar, the most commonly used form of cedar for building purposes, is Western Red Cedar also known as British Columbia Red Cedar. It is grown in Europe and North America. Western Red Cedar is a red-dish-brown non-resinous lightweight softwood, with straight grain and prominent growth rings. It is very durable and often used for vertical cladding garden buildings greenhouses, etc. Cedar contains oil which may impair drying hardening or adhesion of surface coatings. Staining or corrosion of iron and steel fittings, etc., can also be a problem due to the acidic nature of the wood. Staining may also occur if water-based paints are applied. Cedar is a strongly scented wood with a resinous heartwood.

Chestnut is a yellowish-brown hardwood which is extremely resistant to moisture and so very durable. It is mainly grown in Europe. Chestnut can be used for interior and exterior joinery as well as fencing. It does tend to stain when in contact with iron and steel fittings in damp conditions as it contains tannins which could cause problems in terms of appearance. Chestnut used as fencing should be treated with timber preservative. When stained or varnished chestnut joinery can give excellent results.

Douglas Fir is also known as British Colombian Pine or Oregon Pine although botanically it is not a pine at all. It is a softwood which is a light reddish-brown colour with straight grain with prominent 'flame-like' growth rings. Douglas Fir is grown in Europe North America and some parts of New Zealand. It is a very durable wood extremely resistant to moisture and is used for heavy construction work including arches and roof trusses interior and exterior joinery plywood vats and tanks in chemical plants, breweries, distilleries and food processing plants. Staining and corrosion can be a problem in damp conditions when metal fixings are in direct contact with the wood. Douglas Fir gives good results with most decorative finishes although grain raising can be a problem. Wood with a very high resin content should be avoided.

Hemlock is grown in the North and West of America as well as Britain. It is a softwood which is pale brown in colour. Hemlock is not a very durable wood although it is often used for construction and joinery work. It gives good results with varnish or paint systems.

Iroko is a hardwood from Central Africa. It is yellowish brown in colour with an interlocked irregular grain. Iroko is sometimes mistaken for teak although it does have a coarser texture which makes it less attractive than teak. It is a very durable wood with excellent resistance to decay and moisture. It is used for interior and exterior joinery construction work and bench tops. Iroko contains extractives which can retard the drying of solvent-based paints. It can give very good results when stained or varnished. Any open grain should be filled with a suitable coloured wood stopper.

Maple is a hardwood from North America which is very hardwearing although not very durable. It is creamy white in colour. Typical uses are timber flooring in squash courts and furniture. When planed Maple has a distinctive grain figure. It is particularly suitable for staining and varnishing.

Mahogany (African) is an attractive reddish-coloured wood with a highly figured straight grain appearance with some interlocking grain. This may cause problems when staining if the wood has not been sawn correctly. It is used for joinery purposes in the building industry as well as furniture cabinet work and boat building. The best-known timbers of this name are imported from West Africa generally Nigeria or Ghana. African Mahogany is from the khaya family. Makore and Gaboon are also West African Mahoganies.

Mahogany (American) is very similar in appearance to African Mahogany but lighter in shade. Its uses are also very similar. The grain is straight or occasionally interlocking. American Mahogany is generally a reddish-brown colour and is imported from Belize or Brazil. It is particularly suitable for staining.

Meranti comes from South East Asia and is a hardwood. It varies in colour from red to a yellow white. Meranti is not a particularly durable wood and is only used for interior joinery and plywood in the building industry. Meranti has lines of resin ducts that can be troublesome when treated but it does look very effective when stained or varnished.

Oak is widely grown in Europe as well as Japan America and Tasmania. Generally, Oak is a yellowish-brown hardwood with straight grain and a coarse texture. It is a very strong durable wood and very attractive. Oak is often used for interior and exterior joinery flooring gates and is traditionally used for furniture making, panelling and boat building. Oak can be a difficult wood to paint due to long open pores in the grain which need to be filled. It also contains water soluble tannins which can cause staining in damp conditions and may impair the drying and hardening of solvent-based primers when metal work is in direct contact with the wood. Oak is a very attractive wood and gives excellent results when stained or varnished.

Redwood is also known as European Redwood Baltic Redwood Fir Scots Pine Red Pine and Red Deal. It is a softwood varying from a pale yellowish brown to red brown colour. It is mildly resinous and often contains large knots which tend to fall out during sawing. It is used for general construction work, railway sleepers, telegraph poles, etc. The better grades of Redwood are used for joinery work. Redwood is grown in Europe Scandinavia USSR and the Artic Circle. It gives good results with paint varnish and wood stain systems.

Rubberwood small timber blocks laminated to form sections. It should only be used for furniture making, etc. It is not a suitable wood for joinery work.

Sapele is a hardwood medium reddish brown in colour with a pronounced marked stripe in the grain. It is often used as a mahogany although it is harder stronger and more durable. Sapele is grown in West Africa which could explain why it is compared to African Mahogany. Sapele is used for furniture making interior joinery including doors both as a veneer and in solid form. It gives good results when stained or varnished.

Teak is a hardwood golden brown in colour with straight grain which darkens on exposure. It sometimes has dark markings in the grain. Teak is a very decorative wood used extensively for furniture making. It is used for interior and exterior joinery as well as boat building. Teak is grown in India, Burma and Thailand. It is a very durable wood and is extremely resistant to decay and chemical attack. Teak can cause problems when painted due to its oily nature. The surface should be wiped over with white spirit before painting staining or varnishing to remove surface residues. It is particularly suitable for varnishing.

Whitewood, also known as White Deal and Spruce, is a softwood which ranges in colour from white to a pale yellowish brown. It is grown in Europe Scandinavia and Russia and is also referred to as Common or Norway Spruce. Whitewood is not very durable and is only used for some interior joinery flooring and some furniture which is usually supplied untreated. It gives good results with most decorative finishes. It can be used for general construction work. However if used externally, Whitewood must be preservative impregnated the supplier.

REFERENCES

[1] Desch, H. E., and Dinwoodie, J. M. (1996). "Timber: Structure, Properties, Conversion and Use", 7th edn., 306 pp. Macmillan Press, London (ISBN 0-333-60905-0).

[2] Dinwoodie, J. M. (2000). "Timber: Its Nature and Behaviour", 2nd edn., 257 pp. E & FN Spon, London (ISBN 0-419-23580-9).

[3] "Timber for Joinery", 8 pp. (1995). Building Research Establishment, BRE Digest No. 407, Watford, UK.

[4] Fengel, D., and Wegener, G. (1989). "Wood: Chemistry, Ultrastructure, Reactions." Walter de Gruyter, Berlin.

[5] Miles, A. (1978). "Photomicrographs of World Woods." BRE Publications, London (ISBN 0116707542).

[6] Edlin, H. L. (1977). "What Wood Is That?" Stobart, London(ISBN 0854420088).

[7] Farmer, R. H. (1972). "Handbook of Hardwoods", 2nd edn. HMSO, London. (see also 1997 supplement ISBN 1860814107)

[8] "A Handbook of Softwoods." (1957). HMSO, London(ISBN 100114705631).

[9] Bulian, F. (2008). "Verniciare il legno." Hoepli, Milano.

BIBLIOGRAPHY

Speranza, A. (1997), Legno e Umidità, CATAS, Udine.

CHAPTER 3

Raw Materials for Wood Coatings (1) – Film Formers (Binders, Resins and Polymers)

Wood Coatings: Theory and Practice
DOI: 10.1016/B978-0-444-52840-7.00003-5

1. INTRODUCTION

A definition of the term 'coating material' is given by European Standard EN 971-1 (1996) 'Paints and varnishes – Terms and definitions for coating materials': 'a product, in liquid, in paste or powder form, that, when applied to a substrate, forms a film possessing protective, decorative and/or other specific properties'. From this definition, two important aspects regarding coating materials can be considered:

(1) A coating material produces a 'protective and decorative film' after application. Films may be un-pigmented and transparent, as is the case with a varnish or lacquer, or opaque as exemplified by paints. In either case, the essential nature of the product depends on the component that forms a continuous film. Such materials are often simply described as 'film formers' or, in the case of paints as 'binders', since they also have a function of binding or gluing together other components, such as pigment. The film formers used in wood coatings are organic materials, which may be of natural origin (e.g. linseed oil), modified natural (e.g. nitrocellulose) or wholly synthetic (e.g. acrylics, polyurethanes). As such they may be described as resins or polymers and with molecular weights that range from low to very high.

(2) The coating material will be 'applied to a substrate' before forming a solid film. This process can be done only if the coating material is at some stage in a liquid form. The transition state, liquid $\rightarrow$ solid is a fundamental step to the formulation, manufacture, application and behaviour of coating materials. Film formation involves physical and/or chemical changes. Traditionally potential binders are carried in a volatile component in solution or dispersion form. Evaporation of the volatile component, by natural or forced drying, will deposit a film of the binder onto the substrate. Film thickness of the deposited film will depend on the solids content and the spreading rate (see Chapter 9). Volatile components are often described as 'solvents' or 'diluents'. Water is a very special category of volatiles with obvious environmental benefits. Consequently, coatings are often categorised by two broad technology groupings: 'solvent-borne' and 'water-borne'. The legislative implications of this were described in Chapter 1. The term 'water-borne' does not necessarily mean that such coatings are completely solvent-free, they may contain some organic solvent for other purposes, for example, to aid film formation. After evaporation of the volatile components, the film may be in a suitable form for the intended end use. This requires a high molecular weight and has certain constraints as described later. Such films may be described as 'non-convertible'. However in many cases, the molecular weight will be too low to have useful properties. The film must then

be 'converted', while adhering to the substrate, to a higher molecular weight by a process, which is variously described as 'curing', 'cross-linking' or 'thermosetting'. Polymers of relatively low molecular weight may be sufficiently liquid to be applied with little or no solvent present. In this case, conversion is brought about after application by chemical processes which may require addition of heat, catalysts or radiation (Radcure or radiation curing). Thermoplastic polymers may also be applied as a dry powder (Powder Coating) and the required phase change brought about by heating. Subsequently other cross-linking may be initiated by heat or radiation.

From the above discussion, it can be seen that a full understanding of coating film formers and binders involves:

• Chemistry of synthesis or modification
• Technology of delivery (solvent or water-borne, powder, Radcure)
• Physical film formation (drying, coalescence, etc.)
• Chemistry of curing or cross-linking

These topics are dealt with below and in succeeding chapters.

2. CHEMISTRY OF COATINGS

Various types of organic substances have been used in the preparation of coating materials for wood. In ancient times, only natural products such as oils or resins were available. The performance of these products was later improved by thermal or chemical modifications. The continued evolution of industrial chemistry from the 1920s produced many different synthetic film formers such as alkyds, polyurethanes and polyesters which find wide application in all coating sectors including wood. Synthetic resin chemistry can be broadly divided into 'Step Growth' (or condensation) where typically a small molecule such as water is eliminated, and 'Chain Growth' (or simply addition) where repeating units are added to a growing chain. The former is well illustrated by polyesters, and the latter by vinyls and acrylics. In Tables 1 and 2, the most important film former families used for the preparation of wood coatings are listed.

Table 1 Film formers naturally derived

Drying oils and modified drying oils
Natural resins and modified natural resins
Cellulose derivates

Table 2 Film formers synthetically derived

Alkyds
Polyurethanes
Amino resins
Polyesters
Acrylics
Vinyls
Epoxies

2.1. Drying oils and modified drying oils

The name 'oil' derives from the Greek *elaion* = olive tree. From the fruit of this plant a fatty substance, mainly constituted of triglycerides, has been known and widely used for lighting and cooking since ancient times. It is not, however, suited to coating applications. The term 'oil' is also used more generically being given to other classes of substances such as:

• Mineral oils: paraffin blends deriving from petroleum
• Essential oils: natural extractives formed by mixtures of different substances (esters, alcohols, etc.)
• Triglyceride oils

In this section, only selected triglyceride oils, usually derived from vegetable seeds, are considered. Together with some natural resins, oils represent one of the most ancient categories of film formers used for the preparation of wood coating materials. Oils can be used either unmodified, or after thermal and chemical modifications. They represent also an important ingredient for the preparation of the oil-modified polyester resins better known as 'alkyds'.

2.1.1. Oil composition
The glyceride esters are the fundamental constituents of oils. In their natural state, they also contain variable amounts of other components, such as free fatty acids, phospholipids, carbohydrates and sterols; oils are usually refined by chemical treatments to eliminate such substances. Glyceride esters derive from the combination of glycerol (a tri-functional alcohol) with long-chain carboxylic acids called fatty acids (Fig. 1 and Table 3).

Glyceride esters molecules are characterised by three fundamental properties:

• The length of the fatty acid chains
• The presence and the number of double bonds
• The relative position of double bonds (conjugation = two unsaturations separated by a single bond: $-C=C-C=C-$)

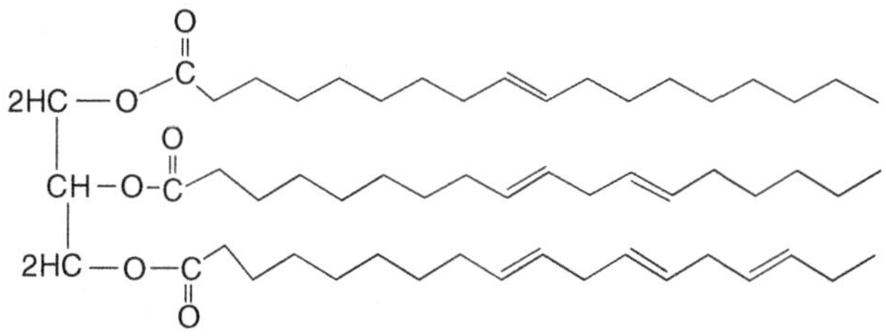

Figure 1 In this example, a typical glyceride ester is illustrated. The tri-functional alcohol (glycerol) is esterified with three different fatty acids. The number of carbon atoms in the fatty acid chain is 26, 28, 30 or 32.

Table 3 Some of the most common fatty acids

Fat acid	Carbon atoms	Double bonds	Conjugation
Myristic	14	0	No
Palmitic	16	0	No
Stearic	18	0	No
Oleic	18	1	No
Linoleic	18	2	No
Linolenic	18	3	No
Eleostearic	18	3	Yes

Natural oils contain different amounts of fatty acids. Formation of a coherent solid film from liquid oil depends on the presence of unsaturated acids (acid containing one or more double bonds). Oils suitable for coatings are usually divided into drying, semi-drying and non-drying groups depending on their composition. This classification is not 'standardised' and some overlapping can be found in the literature (Table 4).

Only drying oils can form a cohesive, solid film when used alone for the preparation of wood coatings. One of the most common and widely used drying oil in wood coatings is linseed oil. It can be used on its own but it is extensively used for the production of air-drying alkyd resins and urethane oils. Tung oil, the most reactive oil, is a traditional ingredient in exterior varnishes and in penetrating treatments for wood, for example, 'Danish Wood Oil'. The semi-drying and the non-drying oils can be used for the modification of alkyds and as plasticisers in nitrocellulose coatings.

2.1.2. Cross-linking mechanism

The formation of a solid protective film from a drying oil, or from an oil-modified polyester (alkyd) involves complex chemical mechanisms known broadly as 'autoxidation'. Although the principal reactions involved in the

Table 4 Classification of some oils, with reference to their acid composition

Type of oil	Name	Saturated acids (%)	Mono-unsaturated acid (oleic) (%)	Di-unsaturated acid (linoleic) (%)	Tri-unsaturated acid (linolenic) (%)	Unsaturated conjugated acids (eleostearic) (%)
Drying	Linseed	10	20	20	50	–
	Tung oil	5	8	4	3	80
Semi-drying	Soya bean	20	20	50	10	–
	Tall oil	10	46	41	3	–
Non-drying	Cottonseed	35	25	40	–	–
	Coconut	91	7	2	–	–

oxidative cross-linking are known, the total mechanism is still not fully established [1]. The entire process can be summarised into three steps: initiation, propagation and termination.

Initiation. The most common theory describing the oil drying involves an initial slow reaction between the atmospheric oxygen and the methylene group adjacent to a double bond. This reaction produces the formation of hydroperoxides. The presence of conjugated double bonds makes the derived free radical stable and in consequence drying reactions are much faster. Several mechanisms have been proposed to explain this difference the most accredited being the initial formation of cyclic peroxides by oxygen addition (Fig. 2).

Propagation (film formation). Hydroperoxides are not stable substances due to the weakness of the oxygen–oxygen bond. The second phase of the drying mechanism involves the decomposition of the initial oxidising products by dissociation of the O–O bonds and the formation of radicals. Radicals are very unstable and reactive species; they produce different molecular arrangements, involving both oxygen and the double bonds. These complex mechanisms lead to the production of other radicals but include also the formation of intermolecular linkage between the oil molecules. It is this cross-linking process that finally produces a solid film (Figs 3 and 4).

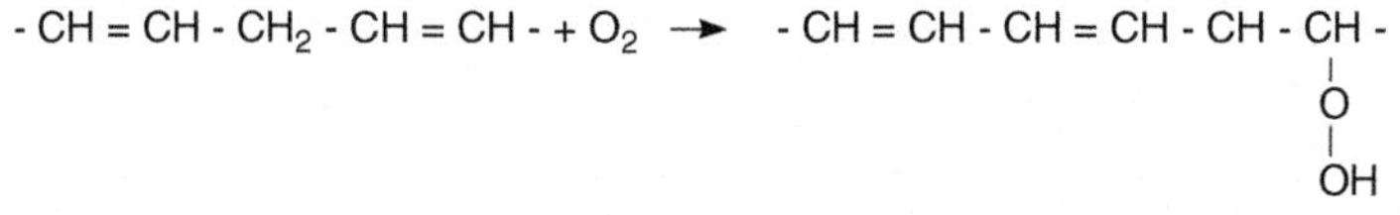

$$- CH = CH - CH_2 - CH = CH - + O_2 \rightarrow - CH = CH - CH = CH - CH - CH -$$
$$\underset{\underset{OH}{|}}{\overset{\overset{|}{O}}{}}$$

Figure 2 Reaction between oxygen and the unconjugated double bond of a fatty acid chain determining the formation of unstable hydroperoxide.

1. ROOH → RO· + HO·
2. ROOH + ROOH → RO· + ROO· + H_2O
3. RO· + RH → ROH + R·
4. HO· + RH → R· + H_2O

Figure 3 Example of the formation of new radicals from hydroperoxides (R = fatty acid chain).

1. R· + R· → R-R
2. R· + RO· → R-O-R
3. RO· + RO· → R-O-O-R

Figure 4 Example of possible linkages between oil molecules (R = fatty acid chain).

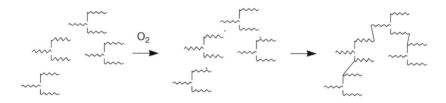

Figure 5 Schematic representation of the triglycerides drying mechanism.

The overall effect of the reactions is that the molecular size of the drying oil increases producing a solid film. During the oxidation process, a number of low molecular weight compounds are also formed, notably ketones and aldehydes. These oxidative by-products are responsible for the typical 'after-odour' of drying oils and their alkyd derivates (Fig. 5).

The uptake of oxygen in the film formation of conjugated oils is relatively lower than in non-conjugated systems. This observation is explained by a theory proposing that the radicals derived from peroxides can also directly involve the conjugated unsaturation producing a chain polymerisation mostly involving carbon–carbon bonding. Peroxides rather than hydroperoxides are involved.

Termination and degradation. The reaction between two radicals leads to the termination of the propagation phase, but the mechanism previously described explains also some important consequences of oil drying and of the behaviour of the coating film so formed:

• There is an induction period before drying during which diffusion and preliminary oxidation occurs.
• The rate of the drying reaction slows down as the cross-linking structure grows due to the difficulty of the oxygen molecules to penetrate inside the coating film. Nevertheless, the cross-linking reactions will slowly continue within the dry coating film even years after application. This leads to embrittlement and is ultimately a cause of failure for exterior coatings.

- The termination reaction (reaction between two radicals) progressively becomes more probable, producing also the formation of peroxide, and ether linkages (molecular weight increase).
- The reaction caused by the polyperoxide linkages (R–O–O–R) and those deriving from the continuous auto-oxidation process lead, during time, to possible depolymerisation and film shrinkage.
- The formation of highly conjugated polymers, absorbing light in the visible region, can derive from certain 'by-products' involving some oils. They can lead to a yellowing of the coating with time. Typically, the more linolenic acid is present (e.g. linseed), the more prone will the dry coating film to yellowing.
- Different reaction paths, according to the nature of catalysts (driers – see below), affect drying behaviour and properties.
- Force drying and stoving produces more carbon–carbon cross-links.

2.1.3. Modified oils

Drying oils may be modified by thermal or chemical treatments. These partly polymerising processes lead to an increase of the molecular weight and consequently fewer cross-links are required to form a coherent film. The drying time of the coating is thereby reduced. Molecular weight growth also produces an increase of the viscosity of the liquid coating material; gloss and hardness are positively influenced by the treatments mentioned above. Thermally treated (heat-polymerised oils in inert atmosphere) without the addition of any other substance, are called 'bodied oils' or 'stand oils'. If the oils are heated and oxidised at the same time by blowing air through the oil they are called blown oils, they are generally faster drying as natural antioxidants are removed. Oils can also be chemically modified to produce film formers with improved properties:

- *Isomerised oils*. These are obtained by heating oil in the presence of certain substances (alkalis and basic solutions) which effect is to increase the extent of conjugation, making them more reactive.
- *Urethane oils*. In this case, oils are previously modified into mono- or diglycerides by heating them in presence of glycerol. The products are then reacted with an isocyanate, producing urethane groups. Due to hydrogen bonding deriving from such modification, coating films become harder and quicker drying.
- *Maleinised oils*. Unsaturated oils can react with maleic anhydride by the 'Ene' or 'Diels–Alder' reaction. Maleinisation increases acid functionality and enables subsequent neutralisation and solubilisation. The latter can be considered as the basic mechanism to produce water dilutable oily film formers. The anhydride group bonded to the oil backbone can be hydrolysed producing cationic groups.

2.2. Natural resins and modified natural resins

The term *resin* is very ancient and is probably of Indo-European origin. It was used to describe different natural substances, found in semi-solid state and used for different purposes. Use of the term 'resin' is often expanded to a generic term for synthetic substances such as phenolic and coumarone resins.

Resins can be considered as a class of polymers being characterised by:

• Solid or semi-solid physical state
• Transparency
• Amorphous morphology
• Solubility in organic solvents

Natural resins can be of vegetable or animal origin. The former may be oleoresins (e.g. Pine Rosin) or hard resins such as 'Damar' which is used in spirit varnishes. Resins such as Congo and Copal are dug from the ground as fossilised exudations. They are generally formed by mixtures of various organic substances (esters, acids, alkanes, etc.), and during their preparation they are generally subject to purification processes and possible chemical modifications.

The use of natural resin is relatively limited in the current coating market although they still play a role in certain sectors being also used to modify some coating properties.

2.2.1. Shellac
Shellac is an animal resinous secretion produced by an insect called *Coccus lacca* during the infestation of certain trees. The term is a combination of *shell* and *lac*, meaning 'lac in thin plates' as it is frequently commercialised in this form. Lac is presumed to derive from the Persian lakh, of Sanskrit origin, meaning 'one hundred thousand' with reference to the number of insects need to produce a small quantity. In ancient times, shellac was mainly used as the source of a characteristic and expensive red pigment. This colorant was separated from the resin and utilised since Roman times. As such it was an alternative to cochineal, which is derived from a closely related insect. The discovery of synthetic colorants (from the end of the nineteenth century) largely superseded the use of natural products for technical and economic reasons. Coating products based on shellac (frequently known as 'French polish') were introduced into Europe in the early 1600s representing the main finishing system for wood furniture until the introduction of the first nitrocellulose film formers in the 1920s. The film former derived from shellac is composed of ethers of long-chain acids, free carboxylic acids, alcohols and hydrocarbons. Shellac can be dissolved in organic solvents such as alcohols and is typically applied on wood surfaces by means a padding technique though spraying and

brushing can also be used. Several applications are needed to obtain a good high build finish. In comparison with other natural resins, shellac-coated surfaces exhibit strong hardness, glossy appearance and resistance to water. Today, shellac is still used especially in the reproduction of furniture where the surface appearance is based on the old traditional procedures of specialised craftsmen. Sometimes, shellac is combined with other film formers (nitrocellulose in particular) to 'balance' the properties of the formulated coating. Unlike some modern finishes, shellac coatings can undergo touch-up repairs. Shellac contains a natural wax, which may resist adhesion of other coating types.

2.2.2. Colophony or rosin oil

It was common in ancient times for materials to be named after their place of origin. Colophony resin takes its name from the Lydian town of Colophon from where these resins came in the past. Colophony is extracted from softwood trees, mainly pine. The separation of the oleo-resin into turpentine and colophony (or rosin) is affected by distillation. Colophony resin is constituted by a mixture of organic acids, mainly abietic, an unsaturated cyclic compound constituted by 20 carbon atoms and containing one carboxylic group. Colophony is soluble in alcohols, turpentine and several other organic solvents. The chemical modification, esterification with polyalcohols, increases the molecular weight of and melting point of colophony with positive effects on the performance of the solidified resin which is used in varnishes. Colophony and it derivates are frequently used as modifiers for nitrocellulose coatings as they increase the coating 'built' (capability to form 'high coating thickness'), gloss and hardness.

2.2.3. Waxes

The word 'wax' derives from the old English word 'weax' = the honey-comb of the beehive, and bees' wax can be considered as the most ancient reference of this category. The various materials named waxes do not form a chemically homogeneous group as they derive from different origin: animal, vegetable or synthetic (Table 5).

Typically, natural waxes are very complex mixtures of different sub-stances. Wax can be considered a sort of general term used to refer to the mixture of long-chain lipophilic compounds being able to form a protec-tive or repellent coating especially towards water. They are frequently used in the wood coating sector representing a special category of film formers which do not form a proper film on wood surfaces conferring water resistance to furniture or flooring surfaces but maintaining the original aspect and structure of the bare wood. Also in combination with oils or resins, they can be dissolved in organic solvents (mainly hydrocarbons) or dispersed into water. Waxes are also used as additives

Table 5 Classification of waxes

Natural waxes		Synthetic waxes	
Fossil	Non-fossil	Partially synthetic	Fully synthetic
Petroleum	Animal	Fatty acid	Polyolefin
– Paraffin wax	waxes	amide	– Polyethylene
– Microcrystalline	– Bees' wax		– Polypropylene
Lignite, peat, montan	Vegetable		Fisher tropsch
– Chemically mods	waxes		Polar synthetic
	– Carnauba		– PTFE

in different coating formulations influencing the appearance (gloss), softness and water repellency.

Bees' wax. Bees' wax (melting point 60–65 °C) is a major constituent of the honeycomb and consists of many different compounds, mainly a variety of long-chain alkanes, acids, esters, polyesters and hydroxy esters.

Its main components are palmitate, palmitoleate, hydroxypalmitate and oleate esters of long-chain alcohols. Nearly 10% of beeswax is based on hentriacontane, whose stability and impermeability to water contribute to the role it plays as a structural component.

Carnauba wax. The leaves of certain palm trees (*Copernicia cerifera, Ceroxylon andicola*) secrete this wax as a protection against hot dry winds. Carnauba wax contains mainly fatty esters, also long-chain alcohols, acids and hydrocarbons. This wax is relatively hard and has a high melting point (around 80 °C) making it very suitable for polishes.

Synthetic waxes. Partially synthetic waxes are made of ethylene glycol diesters or triesters of long-chain fatty acids. Their use has practically superseded the natural derivates due to the availability and reproducible properties. Their melting points can be properly balanced being usually around 60–70 °C. Fully synthetic low molecular weight polyethylene or polypropylene polymers can also be used as waxes.

2.3. Cellulosic film formers

Film formers derived from cellulose can be considered as chemically modified natural products. Cellulose is the main constituent of the cell walls of all vegetables, being a polymer built up from glucose units. Every glucose unit contains three free hydroxyl groups. Cellulose is then a polar substance being with a strong affinity to water. The weight of the cellulose chains is about 300,000–500,000 atomic mass units. Due to its polarity and molecular weight, cellulose is not soluble in any solvent. Cellulose can be used as film former for the preparation of coating material only after a suitable chemical modification of its structure. The initial process is the reduction of its dimensions by hydrolysis reactions. This chemical process produces molecular weight chains of about 50,000–300,000 a.m.u.[1] The second step of the process leading to the preparation of cellulose derivate film formers is the esterification of the free hydroxyl groups with inorganic or organic acids or the etherification with alcohols.

2.3.1. Cellulose esters: CAB and CAP
The esterification of the hydroxyl groups with carboxylic acids (acetic and butyric) produces a resin called cellulose acetate butyrate (CAB).

The presence of the acetate group hardens the resin while the butyrate group improves flexibility, water resistance and compatibility with other resins. Properties and performance can be balanced by adjusting the molecular weight and the proportions of the two acids. A possible variation of CAB is CAP, cellulose acetate propionate. CAB and CAP are film formers greatly soluble in organic solvents, mainly acetates. These resins are widely used in the formulation of solvent-based coatings especially in combination with other 'chemical drying' film formers (e.g. acrylic, radiation-curing coatings, etc.). They provide a number of benefits to different solvent-based coatings including:

- *Drying time reduction.* The great dimension of the polymer chains determines a fast drying time being regulated only by the solvent evaporation.
- *Resistance to discoloration caused by sunlight.* This property mainly depends on the chemical structure where aromatic rings and unsaturations are completely absent.
- *Flow.* The great molecular chain dimensions of the polymeric chains enable the use of such resins as 'flow modifiers'.
- *Levelling improvement.* The solvent evaporation can be balanced to obtain a proper levelling result especially in combination with those film formers (see radiation curing) which drying mechanism is extremely fast.
- *Improve physical properties (hardness).* Also, this property depends greatly on the dimensions of such polymers.
- *Improved transparency of clear coatings.*

More recently, water-dispersible variants of CAB have been introduced to the market.

---

[1] The atomic mass unit (a.m.u.) is defined as the 12th part of the carbon atom mass (carbon atom is constituted by six protons, six electrons and six neutrons).

2.3.2. Cellulose nitrate

The most important esterification process derives from the reaction of hydrolysed cellulose with nitric acid. The final product is called cellulose nitrate. It is usually formed by the action of a mixture of nitric and sulphuric acids. The extent of nitration of the cellulose must be carefully controlled. If cellulose is treated so that practically all of the hydroxyl groups are esterified, the product formed is extremely flammable and dangerous. It is called guncotton and is used in the manufacture of explosives. If the nitration is not carried to completion (with a nitrogen around 11–12%) a stable transparent polymer, cellulose nitrate, soluble in different organic solvents, is formed. The position of the nitrate group in the cellulose chain seems randomly distributed around the –OH terminations (Fig. 6).

The nitric ester of cellulose is usually known as *nitrocellulose* but name is not strictly correct as the groups formed are $C–O–NO_3$ (cellulose nitrate) and not $C–NO_2$ (nitro derivate). The advent of cellulose nitrate coatings that rapidly superseded the traditional products used for the furniture finishing (shellac, oils) is directly related to the invention of the spray guns in the early 1900s. They have represented the main film formers category in the composition of wood coatings until the advent of the polyurethane products, being frequently used also as modifiers of certain properties (flow, drying, appearance) in combination with many other resins. Their cost is low; solubility in alcohols and acetates is very good and the drying time is generally fast being regulated only by the solvents evaporation. In formulation, the solvent balance is often critical and solvents may be categorised as true solvents, co-solvents and diluents. Evaporation rates are selected to avoid incompatibility effects which lead to 'blushing'. Cellulose nitrate usually shows a glass transition around 50 °C. To obtain a coating film with proper adhesion and flexibility, suitable plasticisers (as, e.g. phthalates or alkyd resins) must be used. Cellulose nitrate films are characterised by good hardness and transparency. Their resistance to water is reasonable while conversely they are easily soluble in organic solvents. A potential weakness is poor

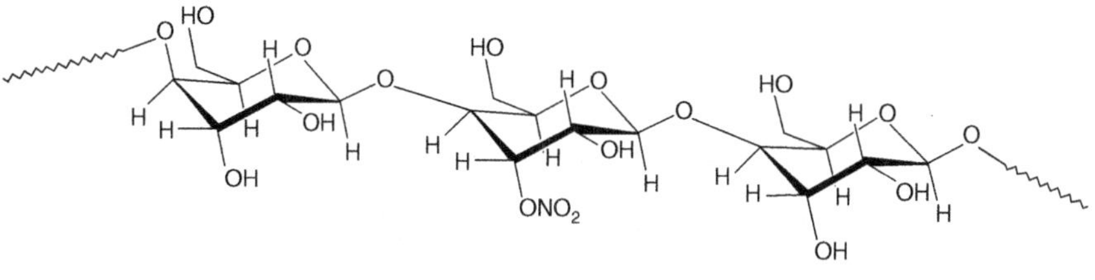

Figure 6   Partially nitrated cellulose chain.

adhesion to the substrate and low performance, mainly due to the absence of a cross-linking process among the polymer chains. The solvent content at the application stage is very high (up to 80% in weight) in consequence of the molecular dimensions of the polymer chains requesting large amounts of solvents to reduce the solution viscosity. As a consequence, cellulose nitrate lacquers have a high VOC which is leading to a rapid decline in their use.

2.3.3. Cellulose ethers

Cellulose can be chemically modified to produce ether derivates. A common example is represented by ethyl cellulose obtained by the etherification of the cellulose hydroxyl groups with ethyl alcohol.

Ethyl cellulose is a colourless polymer soluble in organic solvents being also compatible with many other film formers. Many cellulose ethers are water soluble and used as structuring agents in water-borne coatings.

2.4. Alkyds (oil-modified polyester resins)

Alkyds, considered the first synthetic resins produced for coating application, are the oil-modified ester condensation products of polybasic acids (carboxylic acids containing more than one carboxylic group) and polyalcohols (alcohols containing more than one hydroxyl group). The name derives from the two ingredients mentioned above: alkyd = *alc*ohol + ac*id*. Glycerol and pentaerythrol are the tri-functional alcohols mostly used for the preparation of such film formers. Glycerol is particularly useful as a means to introduce oil into the structure as it can react with triglyceride oils to produce a reactive monoglyceride (Fig. 7).

Phthalic acids or its equivalent anhydride are the 'partners' normally used for the esterification process (Fig. 8).

The orthophthalic is the most common isomer for the preparation of alkyds; isophthalic acid can be used to increase mechanical strength and chemical resistance.

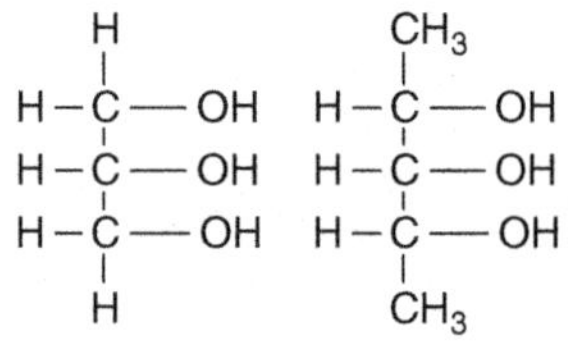

Figure 7   Glycerol (1) and pentaerythrol (2).

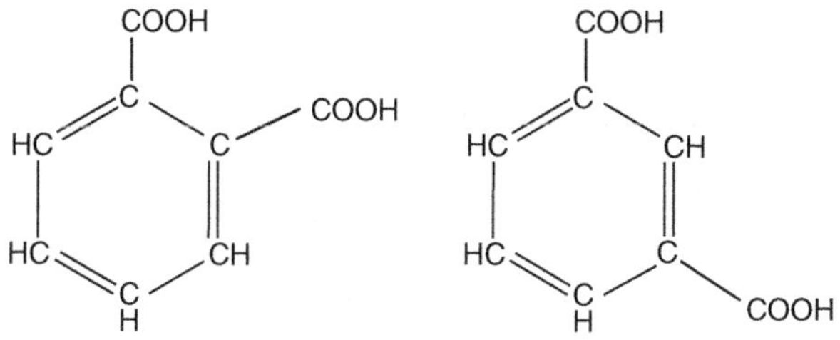

COOH

HC=C
       C—COOH
HC   C
   C   CH
   H
(1)

COOH

HC=C
       CH
HC   C
   C   C
   H   COOH
(2)

Figure 8 Orthophthalic acid (1) and isophthalic acid (2).

CHEMICAL NOTES

Isomers

There are three different 'isomers' of the phthalic acid depending on the relative position of the two carboxylic groups in the aromatic ring. Isomers are two or more compounds having the same molecular formula but different structure (positions of atoms in the molecule).

Anhydrides

The term *anhydride* derives from the Greek *anydros* meaning without (an) and water (Hydror).

Anhydrides are acids deprived of one water molecule. They are important compounds in polymer synthesis as the esterification with polyalcohols takes place without the need to eliminate water.

A characteristic of alkyds is the presence of fatty acids, derived from natural oils, in their molecular structure. The composition of an alkyd can be represented by Fig. 9.

The presence of long carbon chains attached to the alkyd backbone modifies the mechanical properties of the polyester resin, confers solubility in specific solvents and importantly confers the ability to form cross-links through autoxidation as described under 'oils'. Solvents for alkyds are typically aliphatic (e.g. 'White Spirit') or aromatic hydrocarbons. In general, aromatic solvents were displaced by aliphatic ones for reasons of occupational exposure. VOC (Solvent Emission) Directives will further reduce solvent use. However, it is also possible to adapt alkyds to

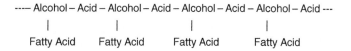

Figure 9 Schematic representation of an alkyd composition.

water-borne technology by emulsification or solubilisation techniques as discussed later. It is also possible to increase the solids content of alkyd solutions in solvent by adjustment of the molecular architecture. On this basis, it may be expected that alkyds will retain their position as a major resin used by the coating industry. Alkyds are classified according the amount of fatty acid reacted (known as 'oil length') and the availability and nature of double bonds derived from the oil component. They may therefore be divided into oxidising and non-oxidising groups. The latter must be reacted with other polymers such as urea or melamine formalde-hyde, to produce films. Oil length represents the weight of triglyceride (formed from fatty acid) as a percentage of the total reacted non-volatile content. Depending on the oil length, alkyds are usually classified into three groups. It is not a formal classification and small differences can be found in literature:

- Long oil: weight of triglyceride > 60%
- Medium oil: weight of triglyceride = 40–60%
- Short oil: weight of triglyceride < 40%

Saturated fatty acids, deriving from non-drying oils such as coconut or hydrogenated castor oil, are mainly used to prepare non-drying (non-oxidising) alkyds which may be used as plasticisers or in combination with specific hardeners (e.g. isocyanates). Unsaturated fatty acids, deriv-ing from drying or semi-drying oils, for example, linseed, safflower, soybean or tall oil, are used to prepare air-drying alkyds (oxidising alkyds). The mechanism by which alkyds dry is very similar to that already described for oils; however, because the molecular weight is higher the amount of cross-linking required is much less and drying times are correspondingly shorter. Only the longer oil alkyds are capable of auto-oxidising at ambient temperatures, but with industrial force drying shorter oil lengths can be used (Fig. 10).

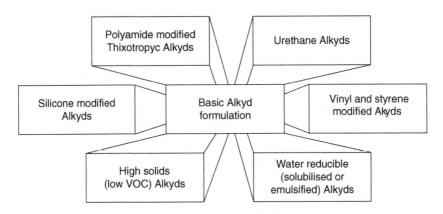

Figure 10 The 'alkyds' family.

Alkyd chemistry has the potential for wide variation as is illustrated in the schematic above and they are widely used as either film formers or modifiers for other resin systems. Within the wood sector, the role of the alkyd will depend on the application with significant difference between building, flooring and furniture. Some typical alkyd uses are summarised below; formulation aspects are covered in Chapter 6.

2.4.1. Drying alkyds
Drying alkyds (medium to long oil) are especially used for the preparation of coating product for architectural wood exposed in exterior conditions, for example, joinery. Alkyds were widely used for interior trim paints but their use will decline as a consequence of solvent legislation.

2.4.2. Polyamide-modified alkyds
Partial reaction of polyamide with alkyds under carefully controlled conditions produces a material with a thixotropic nature (this means this viscosity will fall if the material is subjected to a shearing force over a period of time, but gradually recover when shearing ceases). Polyamide alkyds are widely used in architectural coatings to confer 'non-drip' properties enabling thicker films without sagging.

2.4.3. Urethane-modified alkyds
Residual hydroxyl groups in an alkyd molecule can be reacted with isocyanate to produce urethane linkages and a higher molecular weight. The modified alkyds contain no free isocyanate and used alone, or more commonly blended with other alkyds. The urethane linkages improve hardness, gloss and general mechanical properties. They have good solvent release and hence quicker drying. In wood coatings, they are used as components of architectural paints and varnishes. Urethane chemistry is very versatile (see below) and these modified alkyds should not be confused with solvent-borne or water-borne two-pack materials. Isocyanate/urethane chemistry is also found in polyurethane dispersions (PUDs) described later, which also find wide application in wood coating.

2.4.4. Styrenated and vinyl alkyds
Both styrene and vinyl toluene can be co-polymerised with double bonds from fatty acid chains to improve hardness and drying. However, this introduces more lacquer drying and may lead to recoat problems. Main applications are in the general industrial area.

2.4.5. Silicone alkyds
Hydroxyl or methoxyl functional silicone resins can be reacted with residual hydroxyl groups in alkyds to present a more hydrophobic polymer. Aspects of durability such as gloss retention are improved but at a

significantly increased cost. Paradoxically because chalking is less on exterior weathering, dirt pick-up can be higher (there is no self-cleaning action).

2.4.6. Non-drying alkyds

Alkyds based on non-drying oils are used as plasticisers in combination with 'hard' resins such as cellulose nitrate or amine resins.

Their role is to improve the flexibility of these film formers by reducing their glass transition temperature. Adhesion and gloss is also improved.

2.4.7. High solids alkyds

With the implementation of the VOC legislations, all resin manufacturers have explored the option for higher solids variants and this is also true of alkyds. High solids alkyds are formulated to have a molecular weight (M.Wt) in the range 12,000–20,000 u.m.a (as opposed to the normal range of 40,000–100,000). This may be achieved by various techniques such as chain stopping, and or the use of highly branched fatty acid esters with sufficient residual hydroxyl functionality to react with poly-functional acids. Dendritic resins (branched resins) have also been used. Such approaches can give alkyds with VOC of 200 g/l or even below. Another approach is to replace solvent with a reactive diluent (e.g. a low M.Wt oligomer). However, it must be stressed that high solids resins are not a panacea. They will be more expensive and film formation and cross-linking will proceed differently such that sagging can be a problem. Clearly, it is more difficult to spray a thin controlled film and thus difficult to achieve open-pore effects. In general, the wood coating industry has shown more interest in water-borne, rather than high solids alkyds.

2.5. Isocyanates and polyurethanes

Historically, the reaction between urea and ethanol can produce a compound, similar to an ester, called a urethane (from the initial words of the two components: *urea* + *eth*yl alcohol + -*ane*). Urethanes may also be known as carbamates (Fig. 11).

This white crystalline substance (used in medicine) gives the name to all the class of organic compounds containing a urethane functional

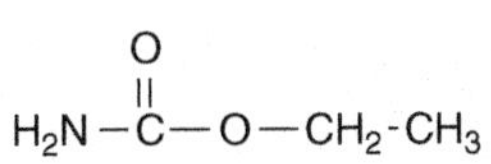

$$H_2N-\overset{\overset{\displaystyle O}{\|}}{C}-O-CH_2\text{-}CH_3$$

Figure 11 The urethane molecule (ethyl carbamate) produced by the reaction of urea with ethyl alcohol.

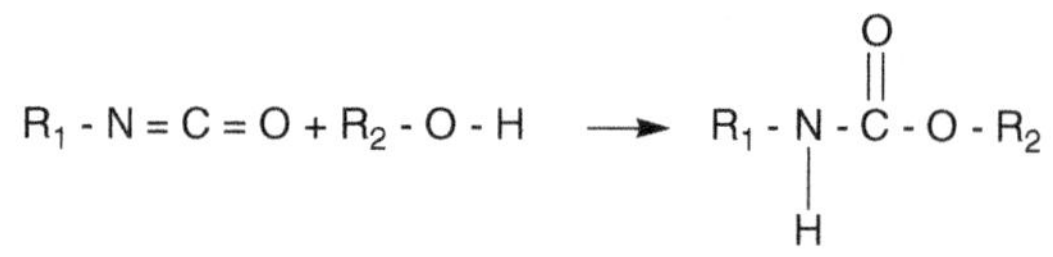

$$R_1 - N = C = O + R_2 - O - H \longrightarrow R_1 - \overset{\displaystyle O}{\underset{\displaystyle H}{\overset{\displaystyle \|}{N} - C - O - R_2}}$$

Figure 12 Urethanes can be usefully produced by a reaction between an isocyanate and an alcohol (or other compounds containing active hydrogen).

group. It was during the 1930s that the German chemist Otto Bayer (1902–1982) laid the foundation for the poly-addition process using poly-functional isocyanates and polyols still used for the preparation of many polyurethane polymers used as plastics, foams and also for the preparation of coating materials (Fig. 12).

From the perspective of the modern coatings industry, *all* polyurethane finishes contain isocyanates or their derivatives.

Isocyanates take the general form R–NCO, where R is aliphatic or aromatic, but they are highly toxic and mostly used in the form of polyisocyanates; even so precautions must be taken when using free polyisocyanate. It should be stressed that isocyanate is used as a modifier or cross-linking agent, rather than the main film former. Thus whereas polyester resins contain predominantly polyester linkages, polyurethane resins need not contain polyurethane resins and even the urethane linkage is not a major component of the dried paint film. Nonetheless, the present of the urethane, which can form intermolecular hydrogen bonds, has a marked effect on properties and 'polyurethanes' are a diverse and useful group of materials.

Binders based on polyurethane can be divided into some broad groupings:

- The isocyanate is fully reacted and the derived binder is used as a 'one-pack' material.
- The isocyanate is kept separate from the other component and used as a two-pack material (often known as 2K).
- The isocyanate is temporarily masked and is used in one-pack form, but must be activated during film formation.

Table 6 illustrates examples of polyurethane chemistry. Technology for delivering polyurethane chemistry includes both solvent and water-borne options. Curing may be brought about at ambient or elevated temperatures.

2.5.1. Properties of polyurethanes

From the above, it can be seen that there are many possible types of resin or binder that may be described as 'polyurethane' with characteristic properties that will include:

Table 6 Examples of polyurethane chemistry

Example	Characteristic features
Urethane alkyd	Modified alkyd curing by autoxidation. Harder and higher gloss than conventional alkyds.
Urethane oil	Hydroxylic oil bodied with diisocyanate but no un-reacted –NCO
Two-pack (2K) urethanes – ambient cure, solvent–borne	Widely used polyol + polyisocyanate combination. Limited pot life.
Blocked urethane – stoving	Isocyanate reactivity is masked as a thermally unstable adduct (e.g. with caprolactam) which is released on stoving.
Moisture-curing polyurethane	Contain an excess of a free isocyanate which reacts with moisture in the air to trigger a reaction via an amine and then urea. May give off carbon dioxide.
Two-pack emulsified systems	Isocyanate is temporarily blocked with an amine; after water and amine have volatilised the –NCO is free to react with polyol.
Water-borne polyurethane dispersions or 'PUDs'	Fully reacted urethane dispersion made by various synthetic routes usually involving dimethylolpropionic acid.
Hydrophobically modified ethylene oxide urethane (HEUR)	A PU block co-polymer used as an associative thickener in water-borne systems.

- Toughness and hardness
- Flexibility
- Abrasion resistance
- Chemical resistance
- Good adhesion
- Fast dry and cure
- Non-yellowing (aliphatic polyisocyanates)

The balance of properties will depend on specific components and the amount of polyisocyanate used. In general, the two-pack and reactive compositions will offer a higher level of performance.

2.5.2. Two-component (or two-pack) polyurethanes (2K)

Two-component polyurethanes are widely used film formers in the high-performance wood coating sector including furniture and flooring. The formation of urethane bonds is based on a chemical reaction being effective during the drying phase of the coating process. The first component contains a polyol (a polymer containing many hydroxyl groups) being often an alkyd or a hydroxyl functional acrylic resin. The second component is a polyisocyanate, known as a 'hardener'; a substance containing two or more isocyanate groups.

Polyisocyanates used for the preparation of polyurethane coatings derive mainly from two basic isocyanates: an aromatic isocyanate, tolylene diisocyanate (TDI) and an aliphatic diisocyanate, hexamethylene diisocyanate (HDI). Other types of polyisocyanate (MDI, IPDI and HMDI) are used elsewhere in the coatings industry.

2.5.2.1. TDI   Toluene diisocyanate is a highly reactive substance. Due to its volatility is also a very dangerous compound causing serious effects to the human breathing system. Consequently, it cannot be used in its simple molecular form at the industrial level, but modified to form a more complex non-volatile polyisocyanate. Three major routes to synthesise polyisocyanates are employed:

(1) Formation of adducts from polyol
(2) Polymerisation to a dimer (uretdione) or trimer (isocyanurate)
(3) Reaction with water to form a 'biuret'

An example of the first reaction is that with a three-functional alcohol (trimethylolpropane). Three molecules of TDI are combined with one molecule of the tri-alcohol. The final adduct is a high molecular weight substance containing three reactive isocyanate groups.

The second strategy is to increase the molecular weight of the TDI, by inducing an internal reaction among three molecules producing an isocyanurate ring where three free isocyanate groups are still present. These two routes are illustrated in Fig. 13.

2.5.2.2. HDI   For the same safety problems noted above, hexamethylene diisocyanate cannot be used in its simple molecular form, and is transformed into a higher molecular weight substance called a biuret. Biuret has a structure related to urea; it is the reaction product of three HDI molecules and one water molecule (Fig. 14).

The reaction producing a polyisocyanate may result in the residual presence of small amounts of the initial ingredients (the diisocyanates) which are not converted to polyisocyanates. The amount of free

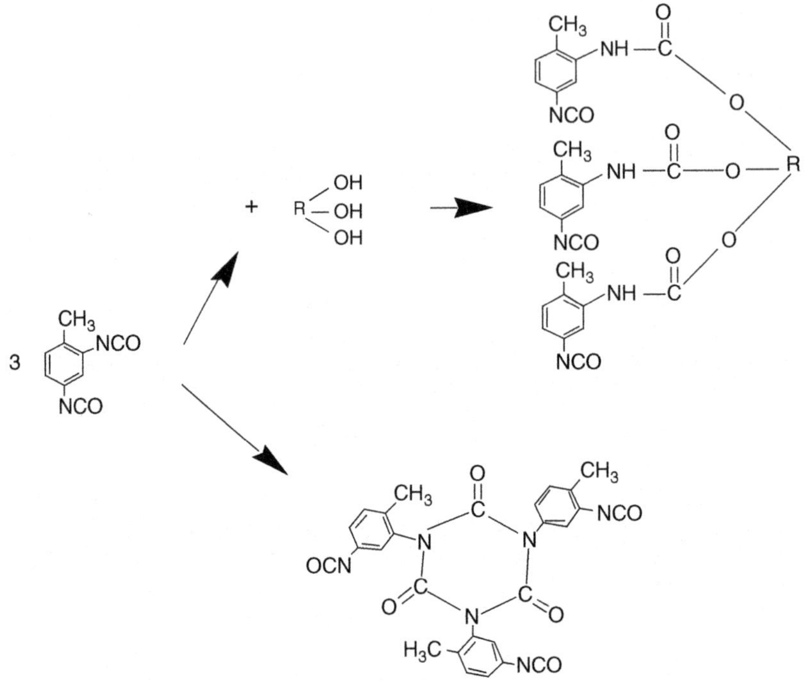

Figure 13 Formation of an isocyanate adduct (diisocyanate + alcohol) and of an isocyanurate ring.

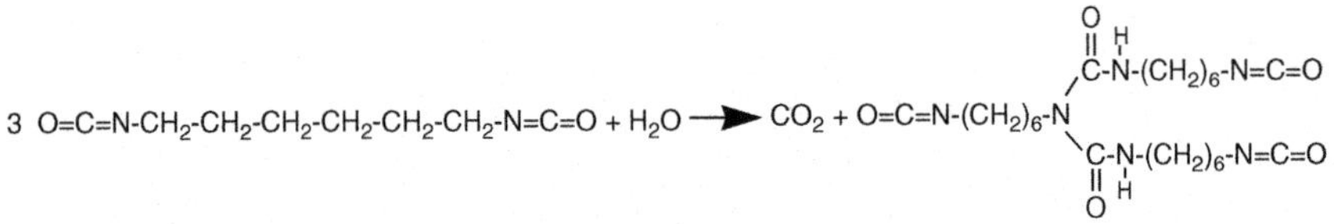
3 O=C=N-CH2-CH2-CH2-CH2-CH2-CH2-N=C=O + H2O
CO2 + O=C=N-(CH2)6-N
C-N-(CH2)6-N=C=O
C-N-(CH2)6-N=C=O

Figure 14 Formation of a biuret (diisocyanate + water).

diisocyanate in the final formulation is an important parameter for health aspects in working environments especially when coatings are applied by spraying and in all cases proper safety precautions must be taken which may include the use of air-fed masks.

For this reason, coating materials based on other film formers have been preferred in certain countries notwithstanding the positive performance attributes mentioned above. The different polyisocyanates strongly influence the application (reaction rate) and the final properties of the polyurethane coating films. Flexibility is higher for polyurethane coatings derived from HDI polyisocyanates while hardness is usually higher with the use of isocyanurates derived from TDI (Table 7).

Table 7 Isocyanates and film properties

Fast cross-linking	Isocyanurate from TD	Hard film
↑	Isocyanurate from TDI and HDI	↓
	Adduct from TDI	
Slow cross-linking	Biuret from HDI	Flexible film

Table 8 Isocyanates and discolouration

Adduct from TDI	Yellowing
Isocyanurate from TDI	
Isocyanurate from TDI and HDI	↓
Biuret from HDI	Not yellowing

Another important property strongly dependent on the polyisocyanate type is the resistance to discoloration (yellowing). Light, and in particular its high energetic wavelengths, can induce a photo-oxidisation of the aromatic rings of TDI leading to the formation of complex quinone compounds. This process leads to the coating film to assume a typical brown–yellow colour (Table 8).

The same phenomenon can arise in the presence of strongly oxidative substances like peroxides. Hydrogen peroxide residues can be present on bare wood as a consequence of bleaching treatments. Peroxides can also be used as 'hardeners' for another film former family (unsaturated polyesters). The contamination of containers or that of the application systems deriving from peroxide residues can bring a yellow discolouration of the polyurethane coating products during either application or final use. The yellowing effect although a major problem for products derived from TDI is effectively absent for coating films produced by the use of biurets.

The properties mentioned above are also modified by the other component of the two-pack polyurethanes, namely the polyol. The use of acrylic resin, free from aromatic rings, is generally preferred when non-yellowing performance is required and other attributes of the polyol can be adjusted to suit the chosen application. In general, isocyanate groups will react with most functional groups containing an active hydrogen atom. This will include alcohols, phenols, amines, amides and water. The reaction product contains an N–H group (e.g. carbamic acid) which is then available for further reaction. In the case of water, this produces an amine and free carbon dioxide. The amine will react with further isocyanate to produce a substituted urea (Fig. 15).

Evolution of CO_2 is utilised to produce foamed polyurethanes but is a problem in coatings and the presence of water, alcohols (primary in

$$R_1 - N = C = O + H_2O \longrightarrow R_1 - NH_2 + CO_2$$

Figure 15 Reaction of an isocyanate with water producing CO_2.

particular) and glycols should be avoided (e.g. in the thinner). Where isocyanate is to be emulsified in water, the –NCO group must be masked, for example, with amine as noted earlier.

2.6. Amino resins (urea and melamine)

Resins containing nitrogen and obtained by the condensation of formaldehyde with amino, imide or amide compounds are classified as *amino resins*. The most representative of this group are urea and melamine resins. Film formers based on urea and melamine resins were introduced into the market during the first decades of the twentieth century, representing a valid alternative to the established cellulose nitrate products, especially when higher performances were required for coated furniture surfaces. Due to concerns over the spraying of isocyanates, acid-catalysed coatings have been considered more favourably in certain areas though it also be noted that permitted levels of free formaldehyde have also been greatly reduced. Conventional amino resins are solvent-borne and therefore subject to VOC legislation. However, high solids and water-borne variants have been developed.

The condensation reaction between urea and formaldehyde produces the urea resin family (Fig. 16); if urea is substituted with melamine, then melamine resins are formed.

The initial condensation of urea (or melamine) and formaldehyde produces highly polar polymers of mono- and dimethylol urea. The reaction is interrupted once the desired molecular weight is reached. A modification is then necessary to make the resin soluble in organic solvents. Etherification of the methylol groups with low molecular weight alcohols (e.g. propyl or butyl) is usually carried out. Due to the presence of three amine groups, melamine has six replaceable hydrogen atoms and methylol melamines can also be alkylated. Fully reacted melamine, hexamethoxymethyl melamine (HMMM) is soluble in water. UF and MF resins are effectively used as cross-linking agents for other resin types

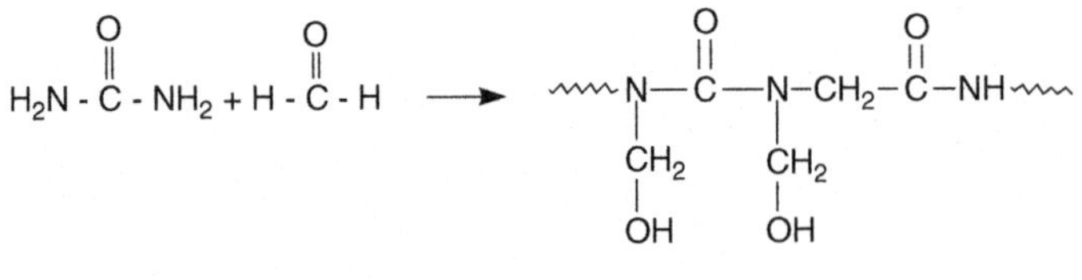

Figure 16 Condensation between urea and formaldehyde.

such as alkyds but will require a catalyst for ambient or force drying conditions. The function of alkyds is to plasticise the coating film.

For wooden substrates, the cross-linking reaction of such resins is effected at room temperature by adding an acid catalyst (a simple inorganic acid such as hydrochloric acid or an organic one as the *para*-toluene sulphonic) just prior to use. For this reason, these resins are also called 'acid cured'.

The resulting applied film cures quite rapidly to a hard, firmly adherent, abrasion resistant film. To balance the drying time, performance and aesthetical quality, the urea and melamine resins are frequently blended with alkyds in similar proportions. Amino resin coatings are valued for high hardness, general resistance and aesthetic appearance. Their high solvent content and the possibility of formaldehyde emission during the cross-linking reaction are the negative points of coatings prepared with conventional forms of amino resins.

2.7. Polyester resins

Every polymer containing ester bonds in the polymer backbone can be defined as a polyester. However in broad terms, it is convenient to divide polyesters for coatings into three groups:

(1) Saturated polyesters
(2) Oil-modified polyesters
(3) Unsaturated polyesters

Saturated polyesters particularly with hydroxyl groups are important components of two-pack urethanes and stoving products used in much industrial coating. Oil-modified polyesters or alkyds have already been described above. Unsaturated polyesters containing double bonds are especially important in auto refinish and as wood finishes. In the coating sector, polymers with the name polyester are derived from the step-growth (condensation) reaction of di- or tri-functional alcohols with di-functional carboxylic acids. Depending on the chemical nature of the initial ingredients, polyesters having a wide range of properties can be prepared (Fig. 17).

The use of aromatic di-functional acids tends to increase hardness in comparison with the same polymer prepared from aliphatic acids. When

$$n\ HO\text{-}R_1\text{-}OH + n\ HO\text{-}\overset{\overset{\displaystyle O}{\|}}{C}\text{-}R_2\text{-}\overset{\overset{\displaystyle O}{\|}}{C}\text{-}OH \longrightarrow HO\text{-}R_1\text{-}O\text{-}\overset{\overset{\displaystyle O}{\|}}{C}\text{-}R_2\text{-}\overset{\overset{\displaystyle O}{\|}}{C}\text{-}O\text{-}R_1\text{-}O\text{-}\overset{\overset{\displaystyle O}{\|}}{C}\text{-}R_2\text{-}\overset{\overset{\displaystyle O}{\|}}{C}\text{-}O\text{-}R_1\text{-}\ etc.$$

Figure 17 Production of a linear polyester chain from the reaction of a di-functional alcohol with a di-functional carboxylic acid.

Table 9 Ingredients for the preparation of polyesters

		Attributes
Glycols	Ethylene	Rigidity
	Propylene	Styrene compatibility
	Dipropylene	Flexibility
	Diethylene	Flexibility
	Neopentyl	UV stability, chemical resistance
Acids	Maleic	Instauration (cross-link sites)
	Fumaric	
	Orthophthalic	Styrene compatibility
	Isophthalic	Mechanical and chemical resistance
	Adipic	Flexibility

glycols with a long carbon chain are used, plasticity and flexibility of the polyester is promoted. Table 9 indicates the broad influences of polyol and polyacid on the attributes of polyesters.

2.7.1. Unsaturated polyesters

The most common source of unsaturation (double bonds) in this type of polyester is conferred by the use of maleic or fumaric acid. Unsaturated polyesters are usually reacted in the presence of a low molecular weight monomer, or oligomer which can take place in the reaction, but also acts as a diluent. The cross-linking reaction of polyesters can be summarised in three different steps:

(1) Activation – formation of radical species
(2) Propagation – reaction of the radicals with double bonds with the formation of new bonds (cross-linking reaction)
(3) Termination – reaction between two radicals

The first step requires some external activators producing radical species. Radicals can derive from a chemical or a photo-chemical reaction (radiation curing or 'Radcure'). Generally, chemical cross-linking applies to polyesters with double bonds in the backbone (e.g. from maleic anhydride), which is in contrast to alkyds where the unsaturation is in side chains. Radiation-curing polymers will often have end unsaturation.

2.7.1.1. Chemical activation Chemical activation is though a redox-initiated system. The coating material is usually supplied as two or three components: the polyester dissolved in a suitable monomer and containing a reducing component (amine or metal soap), and a separate initiator (peroxide or hydroperoxide). The metal soap can also be supplied separately from the polyester. The chemical mechanism of

$$Co^{++} + ROOH \rightarrow RO\cdot + OH^- + Co^{+++}$$
$$Co^{+++} + ROOH \rightarrow ROO\cdot + H^+ + Co^{++}$$

Figure 18   The catalytic role of a cobalt salt.

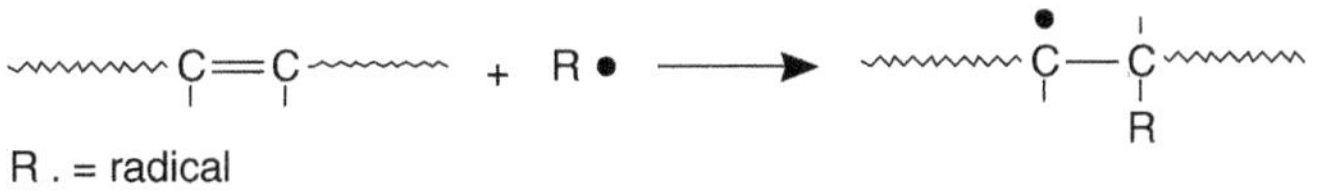

R . = radical

Figure 19   Activation of the polyester chain.

cross-linking can be explained by considering the metal acting as a catalyst, oxidising and reducing continuously as illustrated by Fig. 18 using a cobalt salt and a hydroperoxide.

A typical organic peroxide is methyl ethyl ketone peroxide (MEKP).

Radicals derived from the decomposition of the peroxide are able to react with the double bond, initiating the cross-linking reaction of the film former (Fig. 19).

2.7.1.2. Photo-chemical activation (radiation-curing polyesters – Radcure)
The principle of the photo-chemical activation is based on the use of high-energy electromagnetic waves (UV) promoting the formation of radicals from certain substances called photo-initiators.

The radical species are then able to initiate the cross-linking reaction of the polyester. The technology is often called UV curing and the products, UV polyesters. More details on Radcure technology are given in the paragraph dedicated to acrylic resins.

2.7.1.3. Propagation and termination   Radicals deriving from the ethylene bond scission are then able to interact with double bonds of other resin molecules starting the cross-linking reaction (Fig. 20).

To be able to produce the reaction described above, the radical needs to encounter a double bond on another resin molecule. The probability of such interaction depends on the mobility of the system. As polymers are long and viscous molecules, such reaction is not favoured and the binder is too thick for most application methods. In consequence, polyester film formers are frequently dissolved in a polymerisable monomer such as

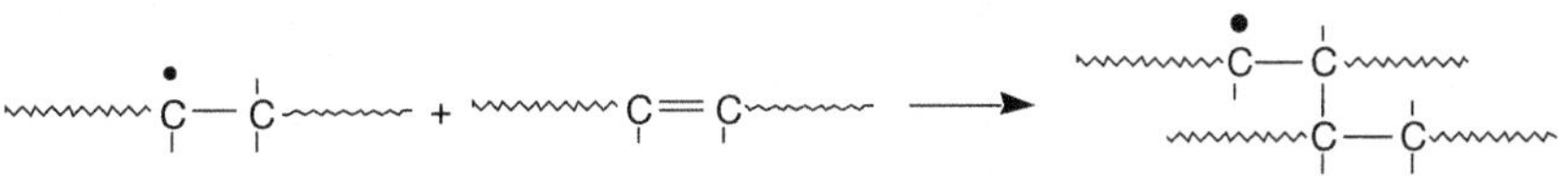

Figure 20   Propagation reaction. Polyester growth.

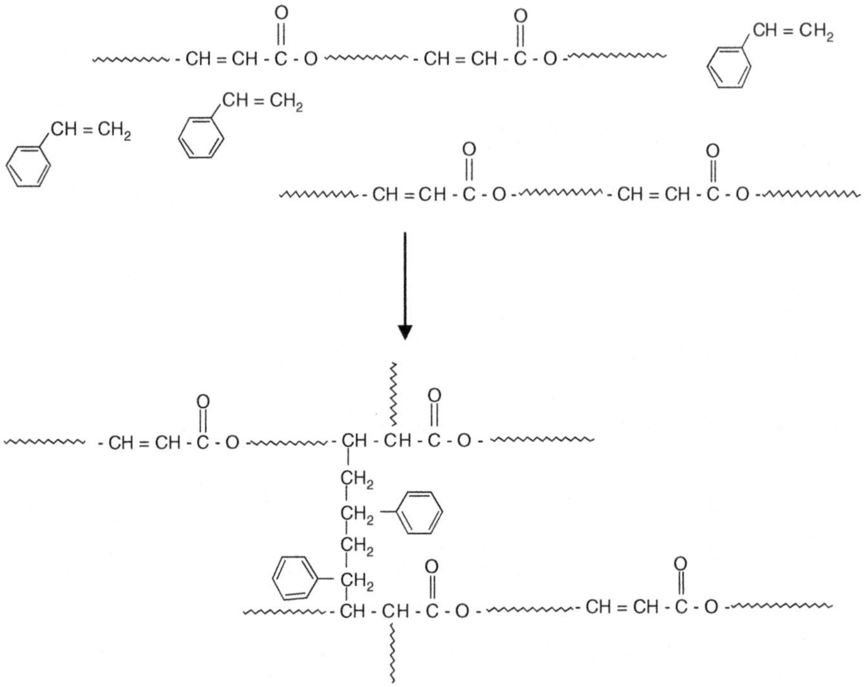

Figure 21 Formation of 'styrene bridges' between two polyester chains.

styrene; the proportion being usually 1 (styrene):2 (polyester). Styrene is a low molecular weight aromatic compound having an ethylene double bond in its structure. Styrene is extremely mobile and consequently the probability of forming chemically activated radical species is very high (Fig. 21).

The cross-linking reactions involving the polyester coatings can be described as the formation of styrene 'bridges' between polymeric chains. Styrene, being a liquid substance in normal conditions, acts both as a solvent (and diluent) and as a partner of the chemical reaction leading to the formation of the coating film.

Styrene and other similar compounds (e.g. vinyl toluene, methyl methacrylate, allyl esters) may then be called 'reactive solvents or diluents' (Table 10).

With reference to the exposure limits related to the use of styrene, polyester containing low amounts of this substance, or even styrene-free polyesters can be prepared. In these formulations, the polyester is dissolved in traditional solvents (e.g. acetates) and the reactivity is balanced acting by reducing the molecular weight and/or using other reactive monomers (e.g. acrylates).

Table 10 Summary of the properties of polyesters

Positive attributes	Some negative aspects
High coating thickness possible	The lifetime of the formulated product is quite short
Glossy finishes	Styrene and other reactive solvents have a relatively low occupational exposure limit
Good chemical and physical resistance	A two-pack (or three-pack) product requires particular attention during the preparation
Fast drying	Air inhibition can occur
Low VOC emissions	

2.7.1.4. Air inhibition Being itself a di-radical molecule, oxygen easily reacts with the radicals present on the top of the polyester layer forming hydroperoxides which dramatically slow down the curing reaction as the result of termination reactions (coupling of two radicals). The drying reaction on the surface is inhibited and the film remains tacky. Three strategies are possible to avoid or even reduce this problem:

(1) The first is the addition of paraffin waxes to the liquid coating. Waxes tend to migrate towards the surface of the film in consequence of their low density and incompatibility. The paraffin layer then generates a partial barrier against oxygen. At the end of the cross-linking reaction, waxes can be removed by means of mechanical treatments (brushing).
(2) The second option is to functionalise the polyester with particular chemical groups able to positively interact with oxygen. Particular substances called allyl ethers can be added, partially replacing styrene (e.g. trimethylolethane mono-allyl ether). Alternatively, polymers are modified incorporating allyl ether groups in the backbone of the polyester chains. Allyl ether groups lead to auto-oxidising reactions via hydroperoxides with reaction mechanisms similar to those previously discussed for the drying oils. Oxygen in combination with such polyesters favours surface drying rather than inhibition. The presence of metal salts, added to activate the peroxides, has a positive effect also towards the allyl groups, catalysing their auto-oxidation (Fig. 22).
(3) A third option is to avoid the reaction with oxygen by the presence of an inert atmosphere during the drying phase. The use of nitrogen, for example, is possible but the high costs involved constrain this option.

The problem of air inhibition is much less with radiation-cured polyesters since the high radical flux allows a much faster rate of reaction and oxygen has less time to dissolve.

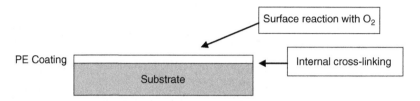

Figure 22   Cross-linking scheme of an allyl-modified polyester.

2.8. Acrylic resins

Acrylic resins are widely used in the wood coating sector being suitable for the preparation of a number of coatings type involving different technologies (solvent, water, Radcure and powder). Acrylics are valued for outstanding properties in areas which include clarity, chemical and physical resistance. In general, the term 'acrylic' refers to the presence of acrylate and methacrylate esters in the polymer structure sometimes with vinyl unsaturated compounds. Vinyl/acrylic chain addition polymerisation is exceptionally versatile and many different polymer architectures are possible (Fig. 23).

The addition reaction involves the double bond present in the chemical structure and the properties of the polymer produced depend mainly on the monomers used, the molecular weight and the possible modification with other substances.

2.8.1. Thermoplastic acrylic resins

Thermoplastic acrylics are usually made by a one-shot solution process and find particular application in automotive and refinish markets as lacquers. They may be internally or externally plasticised and can be blended with nitrocellulose or alkyd in stoving finishes. Molecular weight must be carefully controlled as this influences the orientation of metallic flake pigments. VOC legislation constrains further development. In the wood sector, thermoplastic acrylic resins are mainly derived from a water-borne dispersion route which is described later.

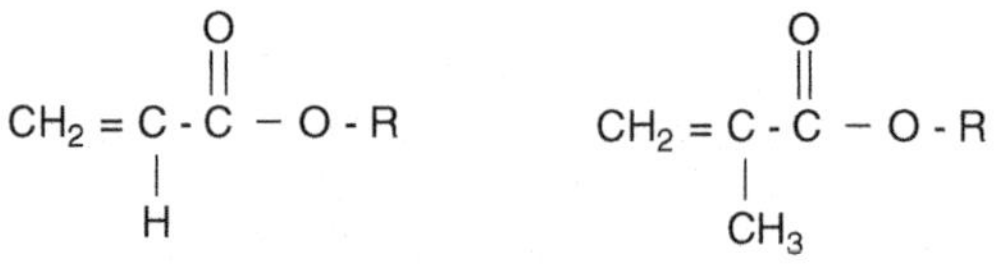

$$
\begin{array}{cc}
\overset{\displaystyle O}{\overset{\displaystyle \|}{CH_2 = C - C}} - O - R & \overset{\displaystyle O}{\overset{\displaystyle \|}{CH_2 = C - C}} - O - R \\
\quad\; | & \quad\; | \\
\quad\; H & \quad\; CH_3 \\
(1) & (2)
\end{array}
$$

R = Methyl (-CH$_3$), Ethyl (-CH$_2$-CH$_3$), Etc.

Figure 23   Acrylate (1) and methacrylate (2) esters.

2.8.2. Thermosetting acrylic resins: Poly-condensation reaction

Thermosetting acrylics are designed to have greater resistance properties than the thermoplastic and find particular application in industrial areas such as white goods. The use of such coatings on wood substrates is less common but may be considered where the substrate is subject to bleaching treatments with oxidising substances like hydrogen peroxide. Cross-linking is brought about with a co-reacting polymer or by an internal self-cross-linking mechanism (see Section 2.8.3).

The initial molecular weight of thermosetting resins is usually lower than the thermoplastic acrylics. The thermosetting acrylic resins contain functionalities (especially hydroxyl groups, e.g. from hydroxyl ethyl acrylate), which can be used in combination with polyisocyanates or other components to give chemical reactions leading to cross-linking. These products can also be regarded as belonging to the polyurethane family. More details on such reactions can be found in the sections dedicated to such film formers. In the treatment of the polyurethane coatings, the acrylic derivates (acrylic resin + polyisocyanate) are distinct from the usual alkyd derivates (alkyds + polyisocyanates) because of certain improved properties of the coating film. The high resistance to yellowing often represents the main reason for the selection of acrylic polyurethane coatings. Aromatic groups are responsible for photo-oxidative reactions leading to the formation of coloured products. Acrylic resins free from aromatics in combination with aliphatic polyisocyanates can be considered as completely non-yellowing products.

2.8.3. Thermosetting acrylic resins: Poly-addition reaction photo-chemically activated (radiation curing)

The principle of the photo-chemical activation is based on the use of high-energy electromagnetic waves to promote the formation of radical species (cationic mechanisms are also possible). The radical species are then able to initiate the cross-linking reaction of the resin. Photo-cross-linking is obtained by means of lamps emitting radiations in the ultra-violet (UV) range. For this reason, the technology is often called UV curing. Another option is electron beam curing (EB) which is more expensive but does not need free radical initiators. EB is also better able to penetrate thick pigmented coatings than UV. UV radiation is able to act on certain substances called photo-initiators added to the coating to produce free radicals. There are two main types of system know as type I and type II, the former cleaves into radicals, while the latter abstract hydrogen from a donor, also known as a synergist. Photo-initiators can be present in the formulated coating or they can be added just before the application. Of particular importance in radiation, curing is the polymerisation of acrylates such as polyester acrylates and urethane acrylates. Unlike the unsaturated polyesters already described, the reactive diluent is an oligomer or pre-polymer rather than styrene. While many radiation-curing polymer systems are used at 100% potential solids content, there are operational situations where some solvent, including water, is used in the formulation to adjust viscosity or reduce the solids content. In wood coatings, radiation curing finds application in areas which include parquet flooring, furniture, wet finishing of doors and other panels, clear coatings on paper and PVC foils.

In contrast to the unsaturated polyesters, used also in combination with peroxides, the acrylated polyesters used in radiation curing have the unsaturated groups at the end of chains rather than in the backbone (Fig. 24).

The resin can then be blended with a poly-functional acrylate as diluent such as tripropyleneglycol diacrylate and others as indicated in Table 11.

The photo-curing resins are a very versatile film formers family allowing the preparation of coating materials with different properties depending on:

- Resin type
- Resin functionality
- Molecular weight
- Resin structure (linear, branched)
- Reactive solvent

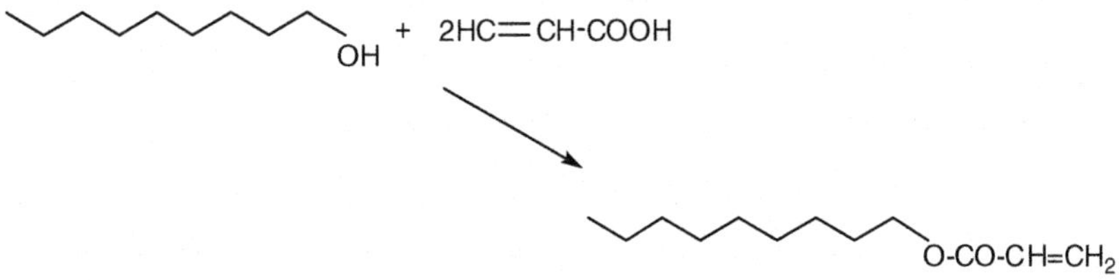

Figure 24   Preparation of an acrylate polyester.

Table 11 Example of acrylates and low molecular weight oligomers

Mono-acrylate	Di-acrylate	Tri-acrylate
2-ethoxyethoxyethyl acrylate	Dipropyleneglycol diacrylate (DPGDA) Tripropyleneglycol diacrylate (TPGDA) Hexanediol diacrylate (HDDA)	Trimethylolpropane triacrylate (TMPTA)

The use of different oligomers can impart different properties to the coating (viscosity, reactivity, hardness, etc.).

A particular category of acrylate resins is known as 'dual cure' as two different curing mechanisms are used. The curing reaction is both a poly-condensation (reaction of the –OH bonds with a polyisocyanate) and a poly-addition (radical reaction promoted by UV light). The drying process under the UV lamps determines a first fast curing. Although the drying is incomplete, it is usually possible to handle the coated elements. The subsequent formation of urethane bonds produces a coating film with the desired properties.

This combination of two drying reactions overcomes some specific limitations of the two single processes (e.g. the incomplete drying of three-dimensional objects, in the case of photo-curing, and the slow drying rate of acrylic resins in combination with aliphatic isocyanates, in the case of condensation reactions).

2.9. Vinyl resins

The vinyl resins derive from the polymerisation of vinyl monomers, and are commonly used to refer to polymers and co-polymers of vinyl chloride. More widely, the term may apply to thermoplastic lacquers bases on vinyli-dene chloride, vinyl acetate and polyvinyl alcohol derivatives. As vinyl monomers contain an ethylene group, a poly-addition reaction can lead to the formation of high molecular weight (polyvinyl) compounds (Fig. 25).

In the wood sector, these lacquers find very little use, but vinyl acetate is an important component of some polymeric dispersions (latexes) as described under Section 3.7.

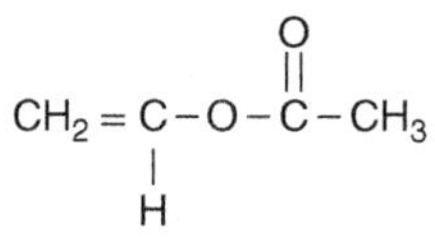

$$CH_2 = \underset{\underset{H}{|}}{C} - O - \overset{\overset{O}{\|}}{C} - CH_3$$

Figure 25   The vinyl acetate monomer.

2.10. Epoxy resins

Epoxy resins are widely renowned for adhesion and chemical resistance but have become to represent a category of film formers used by the wood sector only in recent years when medium-density fibreboard (MDF) sub-strates started to be coated using powder coating technology. The epoxide (or oxirane) group characterises these resins and is formed by a three-atom ring (two carbon and one oxygen atoms). This ring is chemically unstable and easily reacts with many different substances, including carboxyl, hydroxyl, phenol and amine at ambient or relatively low temperatures when suitably catalysed.

The epoxy resins are linear polymers usually containing more than one epoxide group; they are characterised by melting point and epoxide group content. Commercially, a wide range of epoxy resins is available from the reaction of bisphenol-A (diphenylol propane) and epichlorohydrin. The latter is carcinogenic and must be eliminated from the product. Liquid grades of epoxy resins may be chain extended to higher molecular weight with further diphenylol propane. Epoxy resins may be viewed as cross-linking resins, or as polyhydric polyols with the potential to generate further resins. The growth and cross-linking of epoxy-cured resins can be illustrated by the following reactions:

- *Reaction with amines or amides (nucleophiles substances).* The most commonly utilised reaction for polymer growth is that of amines, reacting with the epoxy group according to the following mechanism (Fig. 26).

If a di-ammine is used in combination with a poly-epoxy resin, the condensation reaction leads to the growth of the polymer chains. In the formulation of powder coatings, dicarboxylic anhydrides and dicyandiamide are typically used as hardeners because of their ability of reacting with the epoxy groups without any water of reaction being produced. Reaction with acids other hardeners are based on carboxylic acids or anhydrides. Acid compounds reacting with the epoxy ring and producing a hydroxyl group promote the cationic cross-linking.

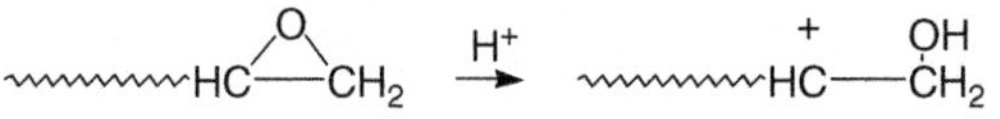

The positive charge derived by the acid leads to a sort of self-cross-linking reaction according to the following general scheme:

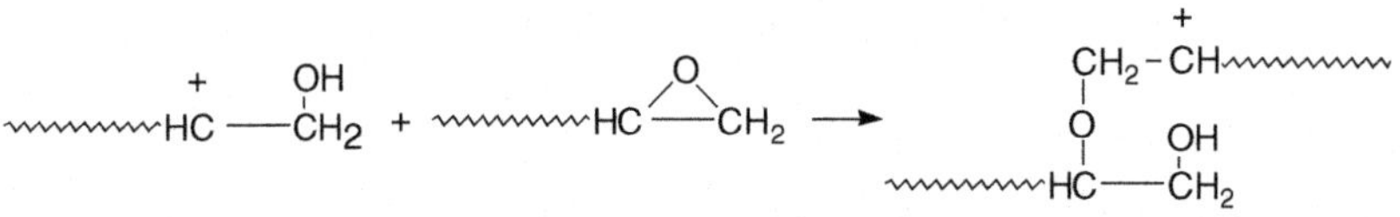

For the promotion of the cationic polymerisation, ferrocene or the 'onum' salts are used (ferrocenium, diazonium, sulphonium, etc.). A recent addition is diaryliodonium salt. When irradiated by the UV light, these salts produced acid substances reacting with the epoxy group.

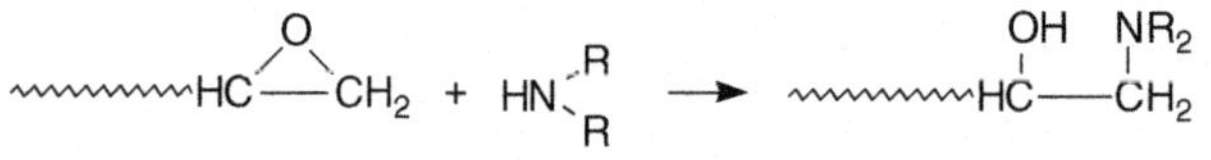

Figure 26 Reaction of an epoxy group with a secondary amine.

The curing of epoxy resins used in the formulation of powder coatings for wood substrates (mainly MDF) is based on the chemical reactions presented above being promoted either by heat or ultra-violet radiations, usually in the presence of a catalyst.

2.11. Epoxy esters

These are equivalent to alkyds with epoxy groups. Both terminal and secondary hydroxyl groups of an epoxy resin can be reacted with fatty acids to produce 'epoxy esters' which may be used as drying finishes or cured with MF resin.

Glycidyl methacrylate enables the introduction of a pendant epoxy group into acrylic resins which improve adhesion and provide a site for further reactions.

3. WATER-BORNE BINDERS AND FILM FORMERS

The preceding sections in this chapter have described the most important categories of chemistry used to prepare potential film formers. In most cases, this has been with the implication that the polymer will be dissolved in solvent, or as is the case with much radiation curing, used at 100% solids. There are clearly environmental and other safety advantages in replacing solvent with water and this has been a major trend in the past decade. Unfortunately, water brings operational disadvantages and in the case of wood this includes increased grain-raising, and problems arising from the high latent heat of water particularly under industrial conditions. These are discussed in more detail in a later chapter. It is broadly true to say that the transition from solvent to water-borne requires retaining the advantages of the resins traditionally carried as solutions in solvent, whilst overcoming a number of operational disadvantages.

Generally speaking, most of the chemistries that have been described above can be translated to water. Rheology of the liquid material will often be different but the dried film will be generically similar. However in some cases, the different physical form of water-borne binders leads to a more open film morphology and differences such as water permeability can arise. The different approached to producing water-borne binders can be summarised as:

- Polymer synthesised in bulk, or solvent, and then transferred water:
  - By solubilisation and neutralisation
  - By emulsification

The former will give a solution or quasi-solution, while the later yields a true liquid in liquid emulsion:

- Polymer synthesised in the presence of water:
  - By emulsion polymerisation
  - Suspension polymerisation
  - Solution polymerisation

The process of *emulsion polymerisation* is the most important of these and despite the name yields a dispersion of polymeric particles. Rheological properties will thus be very different from a solution polymer.

One important category of water-borne urethane resins, namely the polyurethane dispersions, is made by a combination of routes and involves both step and chain growth polymerisation. In all cases, high molecular weight resins (usually as dispersions) can be used without further cross-linking. However for low molecular weight resins and to enhance properties in general, some further cross-linking will be necessary. Thus, it may also be necessary to render the cross-linking agent water soluble or dispersible. Examples of these categories relevant to wood coating will now be given.

3.1. Water-soluble alkyds and polyesters

A route to solubility is to increase the acidity of the resin this may be achieved through the use of trimellitic anhydride or maleinised linseed oil fatty acid. Solubility is greatly enhanced by neutralisation, for example, with dimethylethanolamine, and the degree of neutralisation is an important formulation parameter. It is characteristic of water-soluble resins to show a non-linear dilution response due to molecular associations with water (similar to associative thickeners). Thus, the viscosity dilution curve shows a 'hump' in the intermediate solids range. Water-miscible co-solvents may also be used to adjust viscosity (Fig. 27).

Alkyds and polyesters are prone to hydrolysis on storage and may therefore be diluted prior to use (hence the term *water-thinnable resins*). Acid groups can be cross-linked with a water-soluble compound such as hexamethoxymethyl melamine.

3.2. Emulsified alkyds

Alkyd emulsions are of interest as the basis of external paints or stains for wood. Frequently, there are blended with an acrylic dispersion. In contrast to the solubilisation route described above, alkyd emulsions use predominantly external stabilisers (i.e. suitable surfactants) to form a true emulsion. Manufacture is often by a phase inversion process, that is, to say water is progressively added to the alkyd under high shear mixing, until phase inversion occurs. Phase inversion generally yields a finer particle size

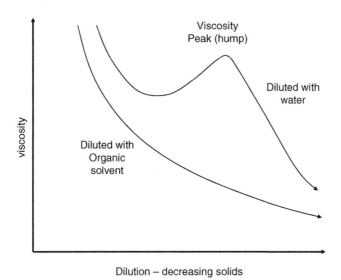

Figure 27   Viscosity behaviour of water-soluble alkyds and polyesters.

product than direct emulsification. Replacing phthalic anhydride with isophthalic acid reduces problems of hydrolysis. Emulsified alkyds must undergo autoxidation to form useful films and will therefore contain driers, though the drier balance will differ from solvent-borne equivalents.

3.3. Water-soluble acrylic resins

These can be prepared by incorporating acrylic or methacrylic acid into the polymer chain and then neutralising with amine. Curing is through the use of HMMM or a water-dispersible blocked isocyanate. These resins are anionic; an alternative cationic route is used for electro-coated resins.

3.4. Water-borne epoxy resins

Both anionic and cationic routes, the later widely used in electro-coating, may make water-borne epoxies. Emulsified or water-dispersed epoxy resins can be cured with amines and find application in anticorrosive primers. Neither is widely used in wood finishing.

3.5. Water-borne two-pack isocyanate systems (urethanes)

There has been considerable development in this area as a route to obtaining good performance (e.g. for wood flooring) without the penalty of high solvent emissions. The non-isocyanate component of the coating system (acrylic or polyester polyols) can be emulsified as described above; however, the isocyanate is problematical as it reacts with water producing carbon dioxide (a useful reaction when making polyurethane foams). Isocyanates can be emulsified into water particularly if rendered hydrophilic with materials such as polyethylene glycol. Viscosity should be kept low and high shear mixing employed. Under ideal circumstances, the bulk of the isocyanate is protected by the formation of a protective skin. Fortunately, the reactivity of isocyanate to other groups is in the order: amine groups > hydroxyl > hydroxyl (water) > carboxyl.

This enables a suitably chosen amine to act as a temporary blocking agent until water has been lost from the film.

3.6. Aqueous polyurethane dispersions

This is a relatively new development in which dispersion in water is followed by chain extension. Many different routes have been patented but one of the more important is a pre-polymerisation process where the polymer contains a carboxyl stabilising group derived from dimethylol-propionic acid (DMPA), which has become an important material in the synthesis of PUDs. After neutralisation and hence dispersion the isocyanate-terminated pre-polymer is chain extended with diamine or polyamine (Fig. 28).

PUDs have significant differences to the emulsion polymer 'latexes' described in the next section. For example, they have a highly water-swollen morphology and even quite hard resins will form a film at room temperature, which has opened up applications for reasonably hardwearing floor finishes in the domestic DIY market. For industrial

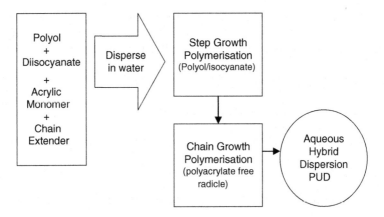

Figure 28   Hybrid polymerisation of acrylic and urethane. (*Note*: This is not co-polymerisation, or a blend, but more an intimate entanglement though there may be some grafting between urethane and acrylic chains.)

applications, further cross-linking is possible with normal external cross-linking agents. Alternatives include fatty acid modification to enable autoxidation and the azomethine reaction. Unsaturated groups will allow radiation-curing options.

PUDs may also be used as a 'seed' to be followed by emulsion polymerisation using acrylic monomers to give true hybrids; they may also be blended with acrylic resins to reduce costs.

3.7. Emulsion polymerisation

The process called 'emulsion polymerisation' produces the largest volume of paint binders used in the coatings industry, due to the large size of the decorative architectural market. Such binders are widely used for domestic wood coating, for industrial joinery but to a lesser extent in other industrial wood areas such as furniture (Fig. 29).

Nomenclature in this area can be confusing. Monomer can be emulsified in water quite readily to produce a true liquid in liquid emulsion. Addition of monomer-soluble initiator to the emulsion causes addition polymerisation and the production of a coarse suspension of particles, which reflect the size of the original emulsion droplets. This process is known as suspension polymerisation and is not much used by the coating industry. If a water-soluble initiator is added in the presence of

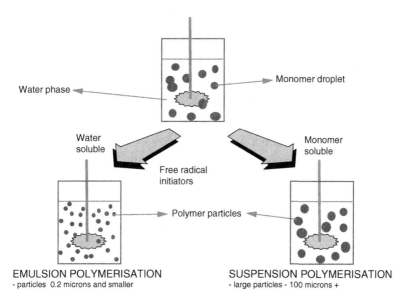

Figure 29 Comparison of emulsion and suspension polymerisation processes. The former produces finer particles.

a suitable surfactant, an entirely different process occurs. New much smaller particles are formed within the micelles created by the surfactant. Monomer diffuses through the water phase and a new high molecular weight dispersion is formed. Further monomer may be added as a 'feed' process. The new polymeric particles are much smaller than the original emulsion. The overall mechanism is complex and will depend on the monomer used. It becomes more complex with co-polymers and ter-polymers with considerable scope to change the morphology of the particle according to the conditions and sequence of monomer addition (see [2]).

Although the process described above produces a polymeric particle (i.e. a dispersion), the paints derived from this process are sometimes called 'Emulsion Paints'. This is technically incorrect particularly in view of other true emulsions such as the alkyd emulsions referred to earlier. Another term in common use is that of 'latex' or 'latex paints'. In this case, the term is derived from naturally occurring rubber latex, and the fact that an early application of emulsion polymerisation was to synthesise rubber.

Polyvinyl acetate (PVA) latexes were introduced into the market in the late 1930s, primarily as adhesives. Styrene butadiene 'latexes' were used for interior finishes from around 1946 and with continued development styrene, acrylic and vinyl products become increasingly common from the late 1950s onwards.

3.8. Composition of water-borne dispersions

The properties of polymers depend on many factors including molecular weight, polymer backbone flexibility and the nature of side chains. Glass transition temperature (T_g) is a useful indicator of mechanical properties and the mobility of polymer segments, and can be estimated from the Fox equation. Knowledge of T_g is useful in designing polymers. To a first approximation, a water-borne dispersion comprises a basic hard monomer chosen for economic and durability reasons, which is co-polymerised with a softer monomer to achieve the overall balance of properties required. Further monomers may be included for specific properties such as adhesion promotion, or to incorporate cross-linking potential.

Some indication of the range of monomers used in emulsion polymerisation is given in Table 12.

Adhesion promoting monomers include:

- Ureido derivatives
- Allyl acetoacetate
- Dimethyl amino methacrylate

Table 12 Monomers used in emulsion polymerisation

Name	T_g of homopolymer	Hardness
Methyl methacrylate	+107	Brittle
Acrylonitrile	+105	Brittle
Styrene	+100	Brittle
Vinyl chloride	+83	Brittle
Vinyl acetate	+30	Hard
Dibutyl maleate	−10	Soft
Ethyl acrylate	−24	Very soft
2-Ethylhexyl acrylate	−50	Very soft
Butyl acrylate	−55	Very soft
Butadiene	−78	Very soft
Ethylene	−125	Very soft

Acrylic and the less-soluble methacrylic acids can enhance stability and provide sites for cross-linking; 1–2% of monomer is incorporated towards the end of the feed process.

Monomer ratios used in typical film formers will be typically 80:20 (vinyl acetate:2-ethylhexyl acrylate) for vinyls but nearer 50:50 for acrylics (e.g. methyl methacrylate:butyl acrylate).

3.8.1. Cross-linking water-borne dispersions

Some of the cross-linking techniques already described can be used with water-borne dispersion provided the necessary functional groups are incorporated. Thus, melamine and isocyanate are possibilities subject to the caveats described earlier. Polyaziridine or a carbodiimide have also been used in two-pack compositions. Methoxysilane provides opportunities for room temperature cross-linking and the use of acetoacetate monomers has also been described.

For one-pack cross-linking, it is necessary to block any reaction until the water has left the film. Application of keto-hydrazide chemistry has been shown to meet this requirement. Adipic bishydrazide is soluble in water and can cross-link keto-functional monomers such as diacetone acrylamide. Hydrazide chemistry may also be applied to reactive stabilisers that have been incorporated during emulsion polymerisation.

3.8.2. Morphology of polymer particles

The term 'morphology' is often used in coatings technology to describe the shape and form of entities in the liquid or dry state; this may include both pigment and polymer components. In the case of polymers, molecular weight is an important parameter, and a given molecular weight may be achieved with different structures and degrees of branching. Highly branched molecules are known as dendrimers and will have different solubility and viscosity characteristics. With polymeric particles made by emulsion polymerisation, there are other opportunities for shape to be controlled. For example by changing the composition of the monomer feed at the seed and growth stage. This can give rise to a core–shell morphology with the outer layer richer in a particular monomer. For example, a carboxylated outer layer imparts novel viscosity control options as the pH is changed. Many other morphologies have been described in the literature to give structured particles which may in some cases lead to non-spherical shapes. The merits of these must be taken on a case-by-case basis, according to the application. The combination of polymeric and pigment particles in the dry paint coating will give rise to different morphologies as the result of packing considerations. This will be very dependent on the size and shape of particles and can influence dry film properties such as opacity and durability.

REFERENCES

[1] Solomon, D. H. (1967). "The Chemistry of Organic Film Formers." John Wiley & Sons, New York.
[2] Gilbert, R. G. (1995). "Emulsion Polymerisation: A Mechanistic Approach", 362 pp. Academic Press, London (ISBN 0-12-283060-1).

BIBLIOGRAPHY

Bulian, F. (2008). "Verniciare il legno." Hoepli, Milano.
Buysens, K., Tielemans, M. and Randoux, Th. (2002). Pv reticolabili per radiazione. Varie tecnologie per diverse applicazioni, Pitture e Vernici n. 19, Milano.
Lambourne, R., and Strivens, T. A., eds. (1999). "Paint and Surface Coatings – Theory and Practice." Woodhead Publishing Ltd, Cambridge.
Nylén, P., and Sunderland, E. (1965). "Modern Surface Coatings." Interscience Publishers, New York.
Paolus, W. (2001). Waterbased radiation curable systems. Principles and application, BASF seminar, Rapallo.
Svane, P., Hacq Y. N., Gard W., and F. Bulian, (1993). Manuale – Rivestimento con polveri del legno e dei pannelli da esso derivati, CATAS, Udine.
Varron, C., Granier V., and Kwee, C. (2003). 2K water-borne PU for furniture coatings, ECJ n. 6., Vincenz, Hannover.
Van Ginkel, M. (2003). I nuovi sviluppi dei polimeri a base acquosa per pv d'uso industriale per legno, Pitture e Vernici, n.2, Milano.

CHAPTER 4

Raw Materials for Wood Coatings (2) – Solvents, Additives and Colorants

1. INTRODUCTION

In Chapter 3, the chemistry and technology of materials, which have the potential to act as a film former and binder in the final coating, were described. Wood, more than most substrates, has a requirement for transparent coatings but even so to be a practical coating the film former will normally only be effective when other 'additives' are present to modify: curing rates, flow properties, etc. Many coatings are not transparent and will therefore require colouring and opacifying agents in the form of

Wood Coatings: Theory and Practice
DOI: 10.1016/B978-0-444-52840-7.00004-7

pigments. Pigments in turn will often require stabilising and wetting additives. Although powder and radiation curing technologies can be used in 100% solids form, the majority of wood coatings are in a fluid form due to the presence of a liquid, which will be lost during the drying process. In this chapter, the main properties of the raw materials other than the binder are described. The means of bringing them together into a 'formulation' are described in Chapter 5.

2. SOLVENTS AND DILUENTS

Volatile transient liquids when used in coatings are often described as 'solvents' and this description focuses on the ability of a liquid to dissolve something, thus according to EN 971, a solvent is a single liquid or blend of liquids, volatile under specified drying conditions and in which the binder is completely soluble.

In some technologies, such as conventional nitrocellulose, a distinction was made between a solvent and co-solvent with the latter only effective in the presence of the first. A further distinction was made between solvent and diluents, which did not dissolve anything but could influence: fluidity, evaporation rate and other properties. Too much diluent would cause precipitation or cloudiness of the film former. Nowadays, terminology is more blurred because, as described in Chapter 3, many polymers are dispersed rather than dissolved in the volatile liquid, and the question of solvency is a secondary issue that is more relevant to components other than the main film former. It must now be accepted that the liquid part of a coating may be variously described as solvent, diluent, volatiles or thinner in a way that is not always consistent.

Classic 'solvents' are low molecular weight organic substances belonging to two different broad chemical families:

(1) Hydrocarbon solvents:
 - Aliphatic (saturated):
   - Straight chains
   - Branched chains
 - Cyclic
 - Aromatic
(2) Oxygenated solvents:
 - Ketones
 - Esters
 - Alcohols
 - Glycol ethers

To this list should be added water, which is one of the most common volatile liquids in coatings, used as both solvent and diluent. Within the oxygenated group are a number of solvents that are compatible with water and may be used to modify various liquid properties; some of these aqueous co-solvents could be regarded as 'additives'. Additional liquid mixtures that are added to a coating immediately prior to use are often described as thinners and will supplement the role of other liquids, their function is:

- Reduction of the viscosity to allow the use of the application system adopted and to allow better penetration, or wetting, of the substrate
- Some adjustment of surface tension
- Balancing the drying time to allow better coating levelling
- Regulating the coating conductance for electrostatic applications
- Reduction of the solids content to obtain evenly controlled thin films and open pore effects

2.1. Solvent properties

2.1.1. Solvency

Solubility is a physical process where the 'affinity' of the molecules plays the most important role. A general rule is that the similarity in the chemical composition can determine a good affinity and thus solubility ('like dissolves like'). Hydrocarbon solvents, being non-polar substances, are usually suitable to dissolve relatively non-polar film formers like oils or alkyds.

Polar solvents like alcohols are more suitable for dissolving relatively polar film formers such as cellulose nitrate or the urea–formaldehyde resins.

There are a number of theoretical treatments of solubility but the most widely used in the coatings industry is the *solubility parameter*. This can be related to the second law of thermodynamics:

$$\Delta G = \Delta H - T\Delta S,$$

where G is the free energy, H is the enthalpy (heat of vaporisation), T is the temperature and S is the entropy.

Essentially something can be expected to dissolve if there is a decrease in free energy. Since molecules will become more mixed on dissolution, there will be an increase in entropy and free energy should decrease, hence solubility is favoured. However, it can be resisted by the enthalpic term. Hildebrand defined a *solubility parameter* to estimate the enthalpic term. The solubility parameter is the square root of the cohesive energy density and is a measure of the molecular attraction:

$$\partial = (E/V)^{1/2},$$

where ∂ is the solubility parameter, E is the cohesive energy/mole and V is the molar volume.

The concept was further developed by Hansen and others, to account for contributions from dispersion, polar and hydrogen bonding molecular forces. Three-dimensional solubility maps may be used to determine solvent and resin solubility. For an introduction, see Ho and Glinka [1].

2.1.2. Viscosity reduction

The amount of solvents (or thinners) added to a coating material, in order to obtain the required viscosity for application purposes for a solution polymer, will vary depending on the chemical nature of the solvents used.

This effect derives from two properties:

(1) The different viscosities of the solvents themselves
(2) The interactions of the solvent with the film former (solubility parameter)

Resin solutions will generally show an upwardly curved increase in viscosity as a function of resin concentration. Paradoxically, the rate of increase is greater for a 'good' solvent as the molecules are fully extended into the solvent. In a 'poor' solvent, the chains are less extended, the mixture is more like dispersion and viscosity is lower. A true dispersion has a viscosity that is less dependent on dilution. Aqueous solutions may show anomalous viscosity–dilution relationships due to association and over a limited range the viscosity might actually rise on dilution.

2.1.3. Evaporation rate

Evaporation rate is an important parameter as it strongly influences the application and appearance of coating materials. Productivity usually requires relatively fast drying speeds but flow characteristics must also be controlled to enable both good substrate wetting and levelling of the film. This can be a difficult compromise and requires an understanding of the rate of viscosity increase relative to the solvent lost; it is of course the reverse of the dilution curve. Dispersion polymers will show a rapid increase in viscosity above around 50% solids as particles come into close proximity. As a consequence, flow and levelling properties are poor. In the case of water-borne dispersions, it is often necessary to add slow evaporating water compatible solvents to extend the 'wet-edge' time.

The volatility of solvents is frequently associated with their boiling point and vapour pressure but modified by resin interactions.

Boiling point and vapour pressure are also used as the key parameter to classify an organic compound as volatile (VOC). This classification is important when environmental constraints on the use or emission of such substances are defined by legislation (see Chapter 1).

Table 1 Boiling point, vapour pressure and evaporation rate of some common solvents

	Boiling point (°C)	Vapour pressure at 20 °C (kPa)	Evaporation rate n-butyl acetate = 1
Acetone	56.1	25.2	7.7
MEK	79.6	9.4	4.6
Ethyl acetate	77.1	9.7	4.1
Toluene	110.6	2.9	2
Isobutyl acetate	118.0	1.7	1.4
n-Butyl acetate	126.1	1.3	1
Cyclohexanone	155.6	0.9	0.3
Butyl glycol	170.6	0.08	0.1

Note. The values reported above were taken from different publications. They could differ from others depending on the solvent purity.

Evaporation rate is strongly influenced by temperature and air-flow rates; for practical reasons it is usually measured relative to a known standard.

Two solvents frequently used as reference standards are diethyl ether and n-butyl acetate (Table 1).

Evaporation rates of solvents in mixtures cannot be predicted from the individual evaporation rates. In a complex system such as a coating material, attractive forces between molecules vary from mixture to mixture being also influenced by the attraction forces with the binder. The evaporation indexes, like simple boiling points, are only a rough guide on which solvent selection can be based.

Water as a liquid carrier faces an additional problem in that water will also be present in the atmosphere as water vapour. Under conditions of high humidity, the evaporation rate of water can be very slow. In industrial situations, humidity-controlled application booths may be required.

2.1.4. Surface tension

Surface tension is caused by the attractive forces (cohesion) between the molecules of the liquid. The molecules within the liquid are attracted equally from all sides but those at the surface are not subject to such interactions in all directions, as they are not fully surrounded by other molecules. Consequently, they cohere more strongly to those molecules adjacent to them on the surface. The surface appears to act like an extremely thin membrane forming a sort of surface 'skin'.

Surface tension is measured in Newton per metre ($N\ m^{-1}$), being equivalent to Joules per square metre ($J\ m^{-2}$). This means that surface tension can also be considered as energy per unit area (surface energy) and in the case of solids it is more appropriate to speak of a surface energy rather than a surface tension.

Surface tension is of great importance to coating formulation and will greatly affect the following properties:

• Wetting and adhesion
• Film formation
• Pigment dispersion
• Application

As the surface tension derives from cohesive forces, it is by definition higher for polar substances than for non-polar ones (Table 2). Temperature reduces the surface tension of liquids.

Liquids with a high surface tension will not wet a low-energy surface; thus water is not a universally effective wetting liquid. This problem may be reduced by the addition of surfactants, which will lower the surface tension and promote wetting. Alternatively, the surface energy of the substrate must be raised.

Surface tension is often measured under equilibrium conditions; such static measurements are not always relevant to dynamic processes such as application and film levelling. Dynamic measurements are more relevant.

Table 2 Polarity and surface tension

Solvent	Surface tension at 25 °C (N m^{-1})	Polarity
Water	0.070	Higher
Ethyl acetate	0.024	$\updownarrow$
n-Hexane	0.018	Lower

2.1.5. Flammability

Almost all the organic solvents are flammable liquids. This means that they easily react with oxygen leading to fast exothermic reactions.

A possible indication of the solvent tendency to ignite derives from a parameter known as the 'flash point'. It is defined as the lowest temperature at which enough vapour form a mixture with air, which can be ignited by a direct flame under carefully specified test conditions.

Flash point is used to classify solvents or liquid coatings for different purposes in relation to potential hazards during application and transport. Regulations will require appropriate labelling.

2.1.6. Electrical conductance and resistance

Resistance is the property of opposing an electrical current and is the reciprocal of conductance. Resistivity is a specific resistance measured in ohm centimetres (Ω cm); resistance is usually expressed in megohms.

The electrical properties of solvents are largely dependent on solvent polarity, but that of liquid coatings is complex, due to the presence of other materials. Electrostatic spraying of liquid paints greatly improves transfer efficiency (see Chapter 9). Droplet size in electrostatic spraying is proportional to surface tension and dielectric constant, being inversely proportional to conductance. Conductance also controls the charging and discharging rate of droplets. A resistivity of between 10^6 and 10^{10} Ω cm is required for most electrostatic equipments and may be adjusted using polar solvents. If this is not possible, other ionic additives may be used (e.g. tetra alkyl ammonium sulphate). Water has a very high conductance and cannot be applied by conventional electrostatic spraying. Special application systems, completely insulated from the can to the gun, should be considered in such cases.

2.1.7. Environmental impact

All the organic substances used as solvents represent a potential safety, health and environmental risk. They may influence workers directly through inhalation, and will accumulate in the atmosphere as pollution.

Limitations for solvent concentration in working environments are published by different organisations. Workplace exposure limits (WELs) are produced by the American Conference of Governmental Industrial Hygienists as guidelines for the protection in the workplace. WELs have replaced a previous terminology of occupational exposure limits (OELs) and threshold limiting values (TLVs). They are obtained from experience and experiment being considered as values to which the majority of the workers can be exposed without negative effects.

The WEL-TWA (time weighted average) is defined as the average concentration during a working period of 8 h. Other limits exist as the WEL-STEL (short-term exposure limits – 15 min) and an upper concentration limit that must not be exceeded at all. These limits, periodically updated, are given together to other information referring the sensitisation and potential carcinogenic effects.

A second negative consequence deriving from the evaporation of solvents is the accumulation and possible interaction with the atmosphere. A well-known problem ascribed to volatile organic substances is the production of ozone at ground level. Ozone, which is however beneficial as a UV screen in the stratosphere, causes serious health problems for growing children, some elderly people, and people with lung conditions such as asthma. Ozone is produced by complex reactions involving organic substances, oxygen and sunlight.

The definition of 'volatile organic compound' is important being often related to environmental legislation (see Chapter 1). Formal definitions are based on one or more the following physical parameters:

- The boiling point
- The vapour pressure
- The tendency to participate to photo-chemical reactions in the atmosphere producing polluting substances (e.g. ozone)

Depending on the reference document considered, different definitions of VOC are possible, for example, between Europe and the USA.

The European directive 1999/13/CE on the limitation of industrial emission of Volatile Organic Compounds defines a volatile organic compound as *(VOC) any organic compound having at 293.15 K (20 °C) a vapour pressure of 0.01 kPa or more.* Another European directive 2004/42/CE, affecting also the wood building sector, define as VOC every organic substance with a boiling point lower than 250 °C at 101.3 kPa.

2.1.8. Odour

The majority of the organic solvents used for the preparation or dilution of coating materials are characterised by particular odours.

An olfactory threshold is defined as the lowest perceived solvent concentration in air, and is different for the various solvents. Human beings show different sensibilities to odours. As the smell of a solvent may be enough to prevent a possible negative exposure, the specific solvent danger is not always connected with the olfactory threshold. Some solvents could be less odorous but highly dangerous and vice versa. Gas chromatography has been used to characterise odour and new techniques involving sensors arrays and neural networks are under development.

2.1.9. Water as a solvent (carrier or diluent)

Water is a unique material and essential to all known life on planet earth. It is the only inorganic liquid to occur naturally and, as a result of its remarkable properties, is found in all three physical states as liquid, solid and vapour. It is cheap, readily available, non-toxic and non-flammable. For these reasons water can truly be said to be in a category of its own, and ideally suited as a carrier for coatings. However as has already been noted, water has a limited capacity to dissolve organic resins and other technological routes including dispersions and emulsions become necessary to exploit waters environmental advantages. Moreover, water has specific disadvantages and may dry too fast or slow according to the conditions; additives remaining in the dry film can cause water sensitivity

Table 3 Comparison between water and *ortho*-xylene

Property	Water	*ortho*-Xylene
Molecular weight	18	106
Boiling point (°C)	100	144
Melting point (°C)	0	−25
Flash point (°C)	None	23
Solubility parameter	49	18
Hydrogen bonding index	39	4.5
Dipole moment	1.8	0.4
Vapour pressure at 25 °C	24	7
Specific heat	4.2	1.7
Heat of vaporisation ($J\ g^{-1}$)	2270	395
Dielectric constant	78	2.4
Thermal conductivity	5.8	1.6
Density ($kg\ dm^{-3}$)	1.0	0.9
Refractive index	1.3	1.5
Ion mobility ($cm\ s^{-1}$)	32×10^{-4}	6×10^{-4}

and water will exacerbate the grain-raising of wood coatings. Some general differences between water and other solvents are illustrated in Table 3.

The properties above show that water is highly associated and very polar; water molecules experience strong hydrogen bonding and this will also affect the viscosity of water-soluble resin solutions. Water undergoes strong hydrophobic interactions that result from entropic causes and as a result will show apparent solubility and viscosity anomalies. Hydrophobic interactions also mean that resins may come out of solution as the temperature is raised. The high dielectric constant results in high ionisation and long-lived ions; thus charge stabilisation is a route to conferring colloidal stability to dispersions in water. Ionic charge is dependent on pH which is therefore an important control parameter in formulation, and may require the presence of a buffer. Water's high latent heat, and heat of vaporisation require additional energy in industrial stoving, and water-borne coatings are sensitive to ambient moisture levels as expressed by the relative humidity. The net result of all the property and environmental differences between water and solvent-borne technologies has resulted in a high acceptance of water-borne in the architectural (decorative) sector. However in industrial wood coating sectors such as furniture, the advantages and disadvantages are more evenly balanced, consequently high solids, powder and radiation curing technologies are also viable options.

RELATIVE HUMIDITY

Water vapour is considered to have reached the saturation in air when the vapour pressure has reached the equilibrium with the liquid water. In these conditions, the relative humidity (R.H.) is considered equal to 100%. The relative humidity of air given at a certain temperature and pressure is expressed by the percentage of water present in air to the maximum possible amount in these conditions (saturation):

$$R.H.(\%) = \frac{\text{partial vapour pressure of water in a sample of air} \times 100}{\text{saturated vapour pressure of water at the same temperature}}.$$

In the following graph, the water content of air in grams per cubic metre is reported at different temperatures for different R.H. The graph shows that the saturation curve considerably increases with temperature. If at 10° the maximum content of water, in the form of humidity, is around $9\,g\,m^{-3}$, it is more than 50 g at 40 °C. Thus, heating air for drying processes involving water-borne coatings produces two major benefits:

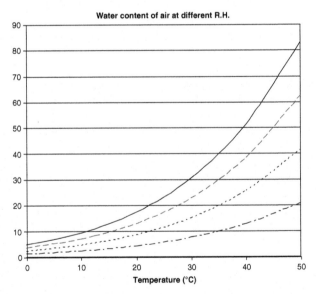

Water content of air at different R.H.

1. Decreasing the relative humidity, allowing more water to evaporate.
2. Increasing the amount of water necessary to reach the saturation (100% R.H.).

Water vapour diffuses into most polymers quite rapidly and will influence mechanical properties including adhesion of the cured film.

3. ADDITIVES

'Additives' (from the Latin *addere* = to add) are described in EN 971 as 'any substance added in small quantities (typically < 1%) to a coating material to improve or modify one or more properties'. The term 'additive' is somewhat derogatory since most coatings would not be viable without modification of the basic properties conferred by the selection of pigment binder and solvent. For this reason, the term 'functional modifier' is arguably better.

Additives can be categorised according to their functional role, or chemistry, as shown in Table 4. Another possibility is to distinguish between those that influence liquid properties, conversion to a film and dry film properties.

From a pragmatic perspective a distinction may also be drawn between additives which:

- are necessary to realise inherent properties (e.g. a pigment dispersant, drier or curing agent)
- confer an additional desirable property (e.g. rheology modifier, biocide, etc.)
- overcome an undesirable side effect (e.g. anti-foam, anti-flooding, etc.)

A problem with the use of additives is that they can interact with each other and are very system specific. 'Circular formulating' is the term sometimes used when additives counteract each other. Thus, a pigment wetting agent may cause foaming and anti-foam is then added which compromises pigment wetting. Clearly, this is to be avoided and this requires knowledge of the chemistry of different modifiers by formulators.

Although additives are used only in small amounts, the global market is valued at around 3 billion Euro, some idea of the relative volume of major categories is shown in Tables 5 and 6.

3.1. Additives affecting the properties of liquid coating materials

3.1.1. Anti-skinning agents
These additives act against the premature surface drying (skin formation) of certain coating materials during storage or production. Skinning is a particular problem for coatings that dry oxidatively such as alkyd resins, and is in character identical to film formation after application. With dispersion paints (latex paints) premature coagulation can be a problem and this can have the nature of a skin. Preventing of skinning is through:

- Anti-oxidants
- Blocking agents
- Solvents
- Drying retarders

Table 4 Additives categorisation

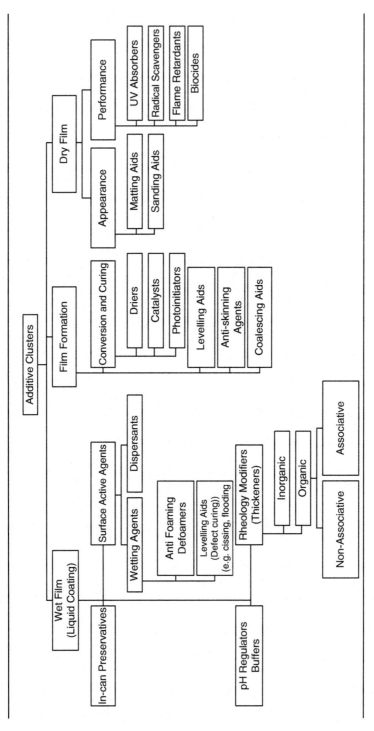

Table 5 The global market of additives

Additive types	Volume (%)	Value (%)
Durability enhancers	8.1	31.3
Rheology modifiers	23.2	23.6
Cure control	43.5	23
Biocides	10	8.8
Surfactants	10.5	7.3
Appearance	17	5.9

Source: IAL Consultancies.

Anti-oxidants must be used with care since they may also retard film curing, they act by competing with the intermediates involved in autoxidation. Anti-oxidants are used at levels between 0.05% and 0.2%, examples of anti-oxidants include hydroquinone, but BHT (2,6-di-*tert*-butyl-4-methoxyphenol) is one of the most common.

Blocking agents are the most common group of anti-skinning agents. The function of these substances is to 'sequestrate' the metals used as catalysts. They form an unstable metal complex. After the application, the metal complex decomposes due to the evaporation of the additive. Metal is then free and able to catalyse the reaction. Most blocking agents are oximes ($>C=N-OH$) and include methyl ethyl ketoxime (MEK), cyclohexanone oxime and acetone oxime. Of these, MEK is the most common and has a fast evaporation rate.

Solvents may prevent skinning by effectively dissolving partially formed films. Slow evaporating solvents are used including dipentene and pine oils.

Drying retarders that slow the loss of water can be useful in waterborne paints and will also extend the 'wet-edge' time. Examples include propylene glycols, and glycol ethers. However, these also contribute the VOC content of the coating and might be partially replaced with a low molecular weight polyethylene glycol.

3.1.2. Surface-active agents

Surface-active agents are an important category of additives, they are amphipathic (from the Greek *amphis* = both and *philia* = friendship) and will have hydrophobic and hydrophilic segments. Normally, they are used to reduce surface tension (e.g. to improve substrate wetting). They play a very important role in pigment dispersion both as wetting agents and as stabilisers. Surface tension and surface tension gradients can also cause problems such as air entrainment, cissing, crawling, fisheyes and other

Table 6 Summary of additives used in the coating industry

Technical field	Functional terms	Chemistry (examples)
Rheology	Structuring agent, thickener, anti-settling or anti-sagging, suspension aids, bodying agents; levelling and flow modifiers	Organoclays (amine mod.), castor oil waxes, metallic soaps, pyrogenic silica, polyamide alkyds, cellulosics, polyacrylates, ethoxylated urethanes, polyacrylamides (associative thickeners)
Electrochemistry	Anti-corrosive (in can, in film, flash corrosion); neutralising and buffering agents (pH)	Chromates, phosphates, amines, zinc, thiocarboxylic acid salts, basic sulphonates; amino hydroxy compounds
Catalysis and cross-linking	Driers Curing agents	Metal soaps (cobalt, zirconium, calcium) of an octoic acid, dibutyl tin laurate; ammonium zirconium carbonate
Free radical modifiers and complexing	Anti-skin Heat stabilisers, flame retardants, anti-oxidants	Oximes and ketoximes Hindered phenols, ceric and antimony oxide; HALS (tetramethylpiperidine)
Surface chemistry and thermodynamics	Surfactants; defoamers, anti-foams, anti-flood and float, wetting aids, emulsifiers, levelling agents, pigment dispersants; colorant acceptors, water repellents	Anionic, cationic, non-ionic and amphoteric agents; silicones; amphipathic polymers; polysiloxanes
Solubility parameters	Coalescing agents; coupling solvents, freeze–thaw stabilisers	Propylene glycol, benzyl alcohol, glycol ethers, esters alcohols, ethylene glycol

Electrostatics	Anti-static; electrostatic spraying	Conductive pigments; polar solvents
Adhesion	Adhesion promoters	Silanes, epoxies
Tribology	Marr and slip agents	Silicone wax; PTFE
Biological and biochemistry	Biocides, bactericides, algicides, anti-fouling	Tributyl tin oxide, formaldehyde, thioridazine-thione, oxazolidines, thiophthalimide
Organoleptic	Reodorants	Esters
Optics	Matting agents, refractive index modifiers	Silica, blanc fixe, ultra-fine TiO$_2$ polymer beads
Electromagnetic electron transfer	Optical brighteners; photo-initiators, synergists, UV absorbers	Benzoxazoles, benzophenone, thioxanthenes, hydroxybenzophenones, tetramethylpiperidine, benzotriazoles, triazines

defects. These effects may be 'cured' by the use of other surfactant-based additives, which must be used with care to avoid antagonistic effects.

Surfactants are usually required to stabilise oil-in-water or water-in-oil emulsions when they will be known as 'emulsifiers' (from the Latin *emulgere* = 'to milk out', from *ex-* 'out' + *mulgere* 'to milk'). Milk is a typical example of emulsion, drops of one hydrophobic liquid (fat) dispersed in water. The selection of surfactants for emulsification is aided by the 'HLB number' which is a measure of the hydrophile (H) lipophile (L) balance (B).

Another important criterion for selecting surfactants is the charge carried by the hydrophile. Surfactants may be cationic, anionic, zwitterionic or non-ionic. In a majority of water-borne paints, an anionic/non-ionic combination is used.

3.1.3. Pigment wetting and dispersing agents

The overall process of pigment dispersion may be considered as three stages:

(1) Wetting
(2) De-agglomeration
(3) Stabilisation

Addition of a powder to a liquid (particularly water) involves considerable air entrainments and the role of the wetting agent is to help displace this air so that mechanical energy can de-agglomerate the powder to the final desired particle size. The dispersion must then be stabilised against re-agglomeration or 'flocculation'. Sometimes the wetting agent will also act as a stabiliser, but in many cases an additional stabiliser will be necessary, often loosely known as a dispersant. Wetting is less of a problem for solvent-borne technologies; furthermore, the main film former may also have a stabilising effect.

There are two major mechanisms of stabilisation:

(1) Charge stabilisation
(2) Steric stabilisation

Charge stabilisation comes from repulsive electrical forces, while steric stabilisation is provided by absorbed macro-molecules to provide a steric thermodynamic barrier. The two mechanisms may be combined.

Often charge stabilisers are small molecules (surfactants) while the steric type is polymeric.

3.1.4. Anti-foaming agents

The possible formation of foam, during application, derives from the inclusion of air due to the mixing of the liquid coatings or the turbulence generated by the application systems used. The term 'defoamer' is used to describe process additives, while 'anti-foams' or 'foam-inhibiting agents' are product additives.

The presence of entrained air can bring to the presence of bubbles or craters in the coating film and is more of a problem with water-based coatings due to the capability of surfactants (such as wetting agents and pigment dispersants) to stabilise air bubbles which might otherwise escape. In the dried film, trapped bubbles become voids with negative effects on appearance and performance. Bubbles which beak the surfaces are known as pinholes.

Reducing the surface tension of the bubble controls foam, but the foam control agent must have a positive entering coefficient, it may be carried on a hydrophobic particle.

Commercial defoamers are typically based on silicon oils and hydrophobic fumed silica. Anti-foams use polyether siloxanes which are more storage stable.

The escape of bubbles is even more difficult for surfaces that are not horizontal and in general it is also wise to take all precautions to prevent air inclusion in the first place. Apart from proper application techniques the entrainment will be affected by viscosity and film thickness. High solids coatings are therefore more vulnerable. In operational terms, two thin coats may be preferred to one thick.

3.1.5. Anti-settling agents

Settlement is a form of sedimentation and is governed by Stoke's law which shows that sedimentation rate increases strongly with particle size and particle density and reduced as the viscosity of the liquid medium is increased:

$$v = kr^2 \Delta \partial / \eta,$$

where v is the settling speed, r is the particle radius, $\Delta \partial$ is the difference between the density of the particle and the liquid medium, η is the viscosity, and k is the constant.

This phenomenon can take place during storage or in any circumstance (even during application) in which the liquid is standing without external mixing.

A good pigment dispersant can reduce sedimentation (and aid redispersion) since the particles are smaller. For dilute suspensions, even good dispersion and stabilisation may not prevent some settlement and paradoxically specific additives may induce a degree of deliberate flocculation. Such additives form a supporting network but must be used with care to prevent loss of properties such as gloss and colour strength. However, the main route to control sedimentation is to have a high viscosity at low shear rates. Rheological additives of many different types can be used (see below). It is important that they do not prevent good flow after application, and this requires shear-thinning behaviour followed by fast recovery to prevent sagging.

3.1.6. Rheological modifiers

Rheology is the science of the deformation and flow of matter. Viscosity is the property of a material to resist deformation by shearing forces; it is defined quantitatively as the shear stress divided by the shear rate. The SI unit of viscosity is the Pascal second (Pa s) but many older units, such as the Poise, are used; some are dependent on the method of measurement. For most coatings, it can be assumed that the relationship between viscosity and shear rate is complex and non-linear; viscosity is thus a function of the shear rate and not a coefficient. It is thus very important to specify the shear rate (units are reciprocal seconds) when describing a viscosity. Low shear rates would be exemplified by the forces exerted by settlement, levelling (due to surface tension), or gravitational sagging. High shear situations would include application and processing. The shear viscosity profile of coatings will change relative to each other as the shear rate is increased and may even cross-over. If viscosity decreases as the shear rate is increased, it is described as shear-thinning or pseudo-plastic. Some coatings, particularly if they contain hard particles, may show shear thickening or dilatancy. If the viscosity decreases when the shear rate is held constant for a period of time, the behaviour is described as thixotropic. Viscosity will recover once the shearing stops, and thixotropy is considered a desirable attribute in some coatings to aid application while resisting sagging.

All coatings will display rheological properties whose value will depend on factors such as the molecular weight of the resin and the phase volume of other material including pigment. On some occasions, the rheology of the basic formulation is suited to the intended application. However in a majority of cases, it will be necessary to add 'rheological additives' that modify the rheology in some way. Areas of application include from manufacture, storage (settlement), application and flow properties during film formation.

Rheological additives are often described as 'thickeners'; they may act directly on the continuous liquid phase, or through various particulate and other interactions. They may be divided into two broad groups for solvent and water-borne technologies within which there are organic and inorganic types. Organic thickeners are usefully further classified as 'associative' and 'non-associative' according to their mode of operation. Associative thickeners and more likely to be used in product where very best flow properties are required. Further information is given in the formulation section (Chapter 5).

3.1.7. Substrate wetting agents

Good substrate wetting is a prerequisite for adhesion, and also for other properties such as spreading and final appearance. Although a coating may be forced into contact with a substrate, this does not result in thermodynamic wetting even though all air may seem to be displaced. Wetting requires the surface tension of the applied liquid to be lower than that of the substrate. This may be achieved, particularly for water-borne coatings by the incorporation of surfactants such as polymethylalkylsiloxanes. This may be a separate functionality to the wetting agents that aid pigment wetting. An alternative to lowering the surface tension of the liquid is to raise the surface energy of the substrate. This is necessary for some plastics and may be achieved through: flame, plasma or chemical treatments.

3.1.8. Coalescening agents

The name coalescent came from the Latin *coalescere* (*com-* 'together' + *alescere* 'to grow up') meaning to fuse together.

Coalescents are temporary plasticisers used in the formulation of water-borne products in order to allow fusion (coalescence) of dispersed resin particles to form a coherent film at an acceptable temperature which may be described as the minimum film-forming temperature (MFFT). Coalescing agents are classed as solvents from the perspective of VOC definitions. Due to increasing restrictions on the solvent content of even water-borne paints, some technologies have been developed which do not require coalescing agents. This has been achieved for example by 'core–shell' morphologies, or the use of co-monomers that enable plasticisation by water itself. Paints derived from such technologies may be described as 'solvent-free water-borne', to distinguish them from more conventional water-borne.

The mechanism of coalescence has been widely studied [2], and takes place in phases that will involve concentration of the film, deformation and inter-diffusion. The latter is seldom complete and hence the dry film will have a different morphology from one that is cast from solution. Higher permeability is often a consequence of this.

The effectiveness of coalescing agents will depend on many factors including: the glass transition temperature (T_g), the structure of the polymeric particle, the affinity for the polymer and water phase and the rate of evaporation from the dry film. Pure hydrocarbons such as white spirit are sometimes used as coalescents, but the majority are polar products such as esters, ketones, ether alcohols and glycols. Many commercial coatings contain a blend of coalescents, which are source from companies specialising in this area. Coalescing agents may also perform other functions in the coating such as to improve wetting and lengthen the period over which the coating can be applied without premature coagulation (wet-edge time). They may also influence compatibility with other ingredients and modify the behaviour of rheological additives.

3.1.9. Biocides
The presence of water and organic compounds as coating ingredients can promote the development of many micro-organisms naturally present in all environments. Biocides are used to control the growth of organisms such as fungi, algae and bacteria. The latter can degrade coatings in storage, especially water-borne coatings, and suitable biocides are often used for in-can preservation. However increasingly, there is concern about health and environmental issues and the use of biocides is covered by increasingly stringent regulations.

In-can preservatives operate through two principle modes: those that are membrane active and those that are electrophilically active (Table 7).

3.1.10. Film preservation agents
Films that are exposed to water, including interior, but more commonly exterior exposure, provide an environment where fungi and algae may flourish. In general water-borne coatings are more vulnerable than solvent-borne, the former are more likely to remain damp (from surfactants and water-soluble additives) and usually are more permeable. Exterior wood provides an additional problem in being a potential nutrient and a reservoir of moisture. Exterior wood coatings of all type are subject to fungal (mould) soiling on the surface and possibly from fungal activity in the wood itself. Blue stain disfigurement from *Aureobasidium pullulans* is a major problem and wood stains are formulated with biocides that may resist this organism.

Table 7 Classification of in-can preservatives

Membrane active	Electrophilically active
Alcohols	Formaldehyde release
Carbanilides	Halogen compounds
Quaternary ammonium	Isothiazolinones
Phenols	Organometallics
Acids	
Biguanides	
Guanidines	

Dry film biocides include:

- Benzimidazoles
- Carbamates
- Dithiocarbamates
- N-Haloalkylthio compounds
- 2-n-Octyl-4-isothiazolin-3-one (OIT)
- Zinc pyrithione
- Diuron

The activity and specificity of these compounds varies greatly and will depend on exposure conditions and other factors [3].

3.1.11. pH regulators and buffers

These substances are added to water-borne coatings in order to regulate the pH depending on the chemical nature of the film formers used. pH is important to control the stability of both polymer and pigment colloidal dispersions. Changing the pH changes the effective charge on a particle (zeta potential = electrostatic potential generated near the surface of the colloidal particles). The pH at which the zeta potential is zero is known as the isoelectric point. Absorption of ionic dispersant depends on the degree of ionisation (also affected by pH) in relation to the zeta potential. pH regulators are usually aliphatic amines (e.g. ethanol amine) or even simple ammonia acting as neutralising agents towards the acid functionalities of the film formers. Ammonia plays also a secondary function as a biocide being antagonist to some bacterial growth.

Buffers are compounds that maintain a required range of pH by acting as a reservoir of acid and base. Specific combinations are used to maintain pH within a desired range. An alkaline buffer solution has a pH greater than 7. Alkaline buffer solutions are commonly made from a weak base and one of its salts.

3.2. Additives controlling the drying (conversion) of coating materials

As already noted, all coatings after application will undergo a liquid–solid phase change. This frequently involves loss of a solvent, and in the case of powder coatings solidification after melting. For some coatings including lacquers, the phase change is sufficient to achieve final properties, and the process is loosely described as 'drying'. However for many coatings, this is insufficient and a further chemical reaction is required to complete the 'drying' process, such coatings may also be described as 'convertible'. The terminologies used to describe the additives that bring about or accelerate conversion are described by a mixture of functional and technical terms, which can vary between technologies. They are often described as 'catalysts', or in the case of UV radiation curing, as 'photo-initiators'. The term 'drier' or 'siccative' is common in the case of oxidative cross-linking.

3.2.1. Driers (oxidative cross-linking)

These substances accelerate the chemical reactions leading to the cross-linking of oil containing paints, especially alkyds. Chemically, driers are soaps of fatty acids in which the metallic cation catalyses oxidation, or acts in some other way to bring about cross-linking. Some materials influence the role of the main catalyst and may be known as promoters or accelerators. Another terminology makes a distinction between 'active' and 'auxiliary' driers.

Autoxidation is a complex process (see Chapter 3) and takes place in several stages. Within the coatings industry, it is common to make a distinction between surface-dry (or touch-dry) and through-dry. Despite some overlap different metals, such as cobalt for surface-dry, and lead for though-dry, are associated with the two stages. Lead, however, has been withdrawn for environmental health reasons, and cobalt is under the threat of legislation (IARC classified cobalt as a possible carcinogen to Humans – Category 2B). This has led to considerable activity on alternatives including manganese and vanadium. Also, resins have been developed which require little or no catalyst. The general role of catalytic driers may be summarised as reported in Table 8.

Table 8 Dryers and drying

Dryers	Drying
Catalytic (top driers, surface driers, oxidative driers) – Co, Mn, V, Ce, Fe	Reactions – Induction – Oxygen diffusion
Cross-linking (through driers, polymerisation driers, coordination driers) – Pb, Zr, La, Nd, Al, Bi, Ba, Sr	– Peroxide formation – Peroxide decomposition – Chain to chain cross-linking
Auxiliary driers (promoters) – Ca, K, Li, Zn	– Through drying – Via –OH and –COOH – Degradation – Weathering and durability
Drier accelerators – 2,2'-bipyridyl	Water problems – Water can act as chain transfer agent
– 1:10 phenanthroline – Chelates	– Hydrolysis – Different drier balance needed

Unfortunately, the autoxidation of oils does not end once the coating has reached a satisfactory dry film; continued oxidation causes undesirable changes such as embrittlement which may ultimately lead to a durability failure.

3.2.2. Catalysts

For isocyanate reactions. As described in Chapter 3, polyisocyanates can react with any active hydrogen including polyols to form polyurethane resins. The reactive isocyanate is added in a minor proportion to the polyol, in line with stoichiometric requirements and as such is often described as a 'curing or cross-linking agent'. However, it might be described more correctly as part of the binder system since it is an integral part of the polyurethane. However, the isocyanate reaction can be catalysed by the urethane product (auto-catalysis) and by metal salts, tertiary amines and organometallic compounds. Two commonly used materials, often used together, are diazobicyclo (2.2.2) octane and dibutyltin dilaurate.

The exact mechanism by which the catalysts operate is complex [4] though it may be broadly assumed that they facilitate proton transfer from the polyol to the isocyanate. Other reactions such as the formation of isocyanurates can also occur.

For unsaturated polyesters. Unsaturated polyester resins are frequently used in combination with monomers such as styrene in reactions that are to take place at room temperatures. Such reactions may be initiated by peroxides.

Peroxides are unstable substances forming radical species in consequence of the breaking of the weak oxygen–oxygen bond.

They are frequently used in combination with accelerators and promoters, to promote radical generation. The most common peroxide is methyl ethyl ketone peroxide (MEK peroxide) used in conjunction with cobalt and salts and dimethylaniline. Cobalt acts as a redox catalyst for decomposition of the peroxide initiator (Fig. 1).

Despite the need for accelerators, it may also be necessary to modify the rate of reaction (e.g. to control the working pot life). Retarders are typically solutions of hydroquinone or catechol: these chemical compounds interact

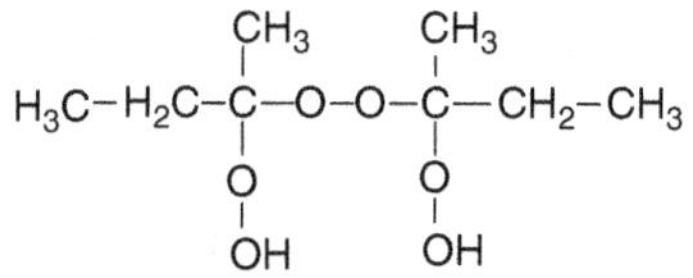

Figure 1 The MEKP molecule.

with the poly-addition reaction forming relatively stable radicals slowing down the cross-linking reaction.

3.2.3. Photo-initiators

One of the processes used to cure coating materials is to promote a poly-addition reaction among unsaturated linear polymers or macro-molecules using UV light. The photo-curing process requires a source of electromagnetic radiation and the presence of particular substances called photo-initiators.

Photo-initiators absorb the electromagnetic energy, at specific wavelengths. This absorption produces fragmentation of the photo-initiator molecules forming free radicals. These radicals are then able to initiate the cross-linking reaction with the film former.

The energy of many chemical bonds is around 5–10 eV thus, considering that also light is a form of energy, the corresponding wavelengths ($E = hc$/wavelength) are those between 200 and 360 nm, belonging to the ultra-violet range.

Radical formation can take place according to different mechanisms:

- *Alpha cleavage of unimolecular photo-initiators.* The photo-initiators belonging to this class are frequently aromatic ketones. They generate free radicals in one single step after the absorption of the UV radiation; however, further cleavage may take place according to the conditions and at higher temperatures more reactive species are produced. The chemical bond involved in the homolytic breaking is that adjacent to the carbonyl group C=O (it is the first bond after this group, being in the 'alpha position'). Depending on the nature of the 'R' group, various types of photo-initiators are produced commercially. These include benzoin ethers, benzyl dimethyl ketone, acetophenone derivates, alpha amino ketones and acyl phosphine oxide compounds. The benzoyl radical formed by the reaction is the most effective towards the initiation of the poly-addition reaction. The other part of the molecule imparts specific properties to the photo-initiator (solubility, stability, etc.) but may also undergo further cleavage. An advantage of the acyl phosphine oxides is that they do not contain acetophenone which is a chromophore imparting a yellow colour (Fig. 2).
- *Hydrogen abstraction of bimolecular initiators.* Hydrogen abstraction initiators do not produce directly free radicals by cleavage but require

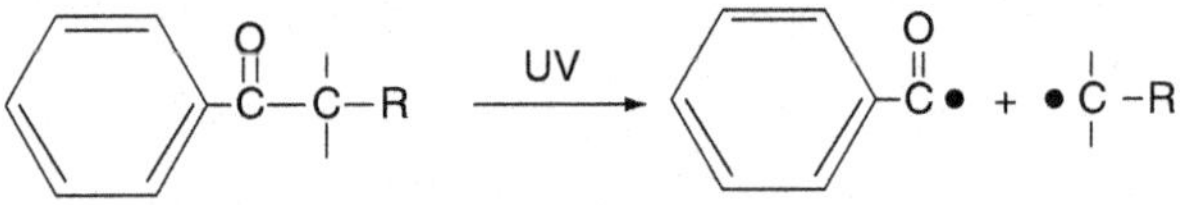

Figure 2 Alpha cleavage.

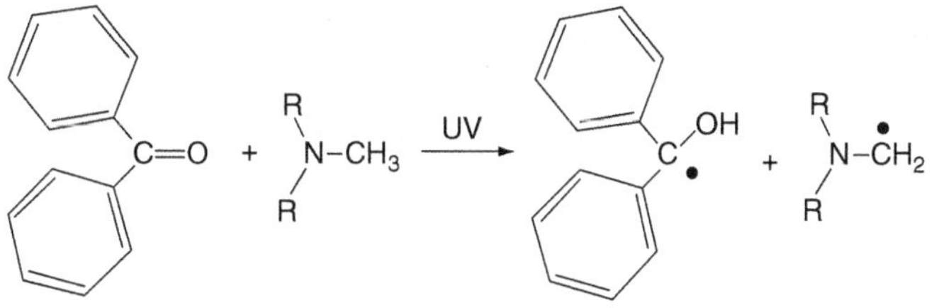

Figure 3   Hydrogen abstraction.

a further step of hydrogen abstraction from a suitable donor. Suitable initiators include benzophenone, xanthone and thioxanthone. The latter are used in the presence of pigments. Widely used hydrogen donors include tertiary amines such as dimethylamino ethanol. One of the useful properties of bimolecular photo-initiators is a lower tendency of the entire process to be inhibited by oxygen. However, a disadvantage is some quenching by vinyl monomers, they cannot be used to photo-initiate styrene reactions (Fig. 3).

• *Cationic photo-initiators.* Cationic initiators are onium salts of strong acids; the strong acid is released by UV light and initiates cationic chain growth. Normally, cationic polymerisation is terminated by water but there is still sufficient reactivity to act as a cross-linker for poly-functional resins. Unlike free radicals, the cations do not self-terminate and reaction can continue even after removal of the UV source. Furthermore, there is no oxygen inhibition. Cationic UV cure is used with cycloaliphatic epoxides, and can be used with vinyl ethers and styrene.

OXYGEN INHIBITION

Oxygen inhibits free radical chain growth mechanisms, by effectively acting as a terminator. This is a major problem with ambient cure of unsaturated polyester resins using peroxide catalysts. The problems is reduced but not always eliminated with UV curing unless the chemistry is modified, for example, with allyl ether co-monomers. Another approach is to carry out the reactions in an inert atmosphere, but this adds operational expenses. Most widely adopted for general purpose resins is the use of semi-crystalline paraffin wax in the formulation. Migration of the wax to the surface both reduces monomer loss and acts as a partial barrier to oxygen. For recoating it may be necessary to remove the wax layer. Paraffin wax is also used to modify other film properties such as water repellence – see below.

3.3. Additives affecting the properties of the coating film

3.3.1. Additives to improve or modify appearance

Two important attributes of coating appearance are colour and gloss or sheen. Colour is primarily imparted by pigments or dyes, and is covered in Section 4. When light is reflected, and refracted, from a smooth surface the source may give a specular and mirror-like reflection with a distinct image. If the surface is less than smooth, the reflection will be diffuse and indistinct. Light reflected at a glancing angle is more intense than at a normal angle and less affected by surface roughness. In general, the light reflected at an angle of 60° from the normal is termed 'gloss' while at 85° or below it is termed 'sheen'. The terminology may be used imprecisely as it is conflated with marketing terms such as 'silk' or 'satin'. Many sheen/gloss profiles are possible and this allows considerable product differentiation. For a full characterisation, the incident and reflected light need to be set and measured over the full 180° range. Results will be dependent on the nature of the equipment used ranging from simple 'gloss metres' to full goniophotometers. Different instruments are needed to quantify 'distinctness of image'. The perceived gloss of a coating will depend on colour and refractive index, and in particular the surface roughness. It is useful to distinguish between micro- and macro-roughness; the former arises from film shrinkage and the presence of sub-micron inclusions, while the latter may arise from application (e.g. brush marks, spray mottle) and flow induced by surface tension gradients. The deliberate inclusion of fine particles is known as 'matting' and the materials as 'matting agents'. Levelling agents are added to improve the flow-out of defects. Gloss and sheen will also be influenced by surface preparation including substrate and inter-coat sanding.

3.3.2. Sanding additives

To obtain the most attractive appearance in wood coating, it is common practice to build the total surface coating from a series of thin layers with inter-coat sanding using various grades of sandpaper. Sanding is also necessary to minimise the effects of grain-raising. Specific additives, frequently metallic soaps such as zinc stearate can act as a lubricant. Zinc stearate is also used as a matting agent and can fill pores.

3.3.3. Matting agents

Inorganic. Inorganic matting agents cover natural and synthetic extender pigments. Extenders may be loosely described as pigments with a low refractive index – see later section under 'Pigments'. They are usually colourless and can only confer low opacity to the film, and if the refractive index is the same as that of the medium they will appear transparent.

This is useful for adjusting the gloss of clear finishes. Natural silicas (diatomaceous earth) have been widely used in wood coatings but some are subject to legislation as potential carcinogens. They are being replaced with synthetic micronised silicas and are available in porous and non-porous forms sometimes with an organic wax coating to prevent settlement and agglomeration.

Organic. Waxes, both synthetic and natural, are useful as matting agent, sometimes in combination with silica. To some extent waxes can concentrate at the surface of a film and may confer tactile effects. Synthetic waxes include polyolefin and polyethylene and may be incorporated as hot blends or cold dispersions.

3.3.4. Levelling agents

The term 'levelling' is used to refer to the spontaneous smoothing of a film after application and is driven primarily by surface tension; the term also refers to the elimination of defects that appear after application (e.g. craters, fisheyes and orange peel). The two main polymers used to aid levelling are polyacrylates and cellulose acetate butyrate CAB, silicones and fluoro-surfactants are also used. CAB is widely used in industrial wood coatings.

3.3.5. UV absorbers

These additives are added to coating materials with the scope of protecting the surface against the energetic ultra-violet radiation of sunlight. The effects of such interaction can be a simple colour change or even a degradation of the substrate and the film itself. The UV absorbers usually act absorbing incident radiation below 400 nm and converting it into heat, a harmless form of energy, which is then dissipated.

Benzophenones, benzotriazoles and phenyl triazines derivates are the most frequently used UV absorbers. The main differences between such additives depend on their absorption in the UV range from which derives their protective effect. Another important property is their stability. A problem for wood substrates is that wood is degraded by visible light at the blue end of the spectrum as well as UV light. This has led to the development of modified UV absorbers, which are red-shifted and improve performance. There is, however, a limit to this approach before the absorbers become too visibly coloured. There remains therefore the problem of protecting wood under transparent coatings from lignin oxidation by visible light and this is addressed by the use of a radical scavenger in combination with the UV absorber.

The following graph shows the absorption spectra of three films of a clear acrylic resin in combination with three different UV absorbers. All the three curves show a complete transparency in the visible range and different absorptions in the UV part of the spectrum (Fig. 4).

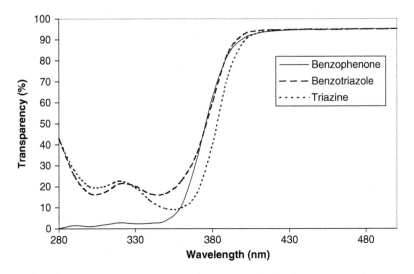

Figure 4 UV-Vis spectra of an acrylic resin with three different UV absorbers.

Another opportunity to protect the wood substrate from the degrading action of UV light is the use of inorganic pigments. As the coating film is made completely opaque, no wavelength is allowed to affect the substrate.

A special case is that of finely divided 'nano-size' transparent oxides (mainly iron, zinc and titanium oxides). Transparent oxides differ from the opaque ones principally in terms of particle size and shape. Such compounds show a relative transparency in the visible range and an intense absorption in the UV range.

Transparent oxides are generally much stable against UV light and heat being frequently used in combination with UV absorbers in the formulation of clear coatings for outdoor use (Fig. 5).

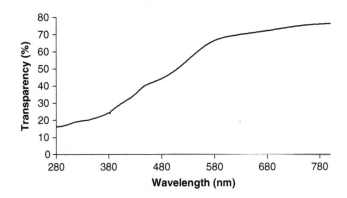

Figure 5 UV-Vis spectra of a transparent iron oxide.

3.3.6. Radical scavengers

As noted above, these additives are frequently added to coating materials for exterior use, to prevent the degradation effects of radical species formed by exposure to sunlight.

Radical scavengers are normally used in combination with UV absorbers as they do not act absorbing UV light, but by the trapping (scavenging) of free radicals. The most widely used class of radical scavengers is represented by the hindered amine light stabilisers (HALS). They form stable products by reacting with radical species generated in the film former. The whole mechanism, being rather complex, regenerates the radical scavengers, which are then able to act continuously against the possible formation of new radicals.

3.3.7. Flame retardants

These substances are added to some coating materials in order to slow, retard or suppress the risk of fire of coated products.

Wood, its derivates and also coating films are organic materials subject to combustion when exposed to direct flames or to high temperatures in presence of oxygen. The presence of free flames and high temperatures are typical of fires and so, due to safety regulations, under some circumstances certain products (e.g. floorings) should be coated with appropriate fire-retarding coatings. National laws regulate this subject in many countries. The retarding effect is given to the coating materials by specific additives and formulating practices.

To better understand their behaviour, it is useful to briefly remember the physical chemical mechanisms involved during the combustion process of a polymeric material:

(1) The initial phase involves the heating of the polymer by an external source.
(2) Heat causes melting and decomposition with formation of volatile compounds (endothermal process).
(3) In the third phase, such decomposition products are oxidised by oxygen; high temperatures, deriving from the surrounding atmosphere, catalyse the reaction.
(4) The fourth phase is the auto-propagation of the combustion process due to the same heat produced by the oxidative reaction (step 3).

Fire-retardant formulation acts against one or more of the above-mentioned steps by physical or chemical methods.

Physical action: protective shield from charring and intumescence. A protective carbon barrier can be generated by the effect of heat on certain additives. The reaction also involves the polymer forming the coating film. This barrier acts against the combustion by the following two properties:

(1) Low thermal conductivity slows down heating of the surface.
(2) The barrier prevents the substrate from direct interaction with oxygen.

Polymers containing hydroxyl groups are first dehydrated leading to the formation of double bonds, which further cross-link to form carbon-rich structures. Phosphoric acid derivatives, and also boric acid may also decompose to structures, which can further char. Charing can be taken a stage further by enabling expansion into a foam or char to enhance thermal insulation. This type of fire prevention is the basis of 'intumescing products'.

Intumescing is a word deriving from the Latin *intumescere* meaning to 'swell up'. When exposed to high temperatures, these coatings expand to protect the substrate from fire. Ammonium phosphate decomposes providing ammonia as an inert expansion gas and phosphoric acid to catalyse decomposition of the carbon source. Resins such as melamine and guanidine are added to melt and bind the foamed char into a coherent mass. Although wood is itself subject to combustion, it can also form a char that provides a margin of safety during a fire.

Physical action: endothermic effects. Another physical strategy is the use of additives generating inert gases or water during their thermal decomposition. The effect is a dilution and cooling of the combustible gases deriving from the polymer decomposition.

The cooling effect can also derive from the endothermic reactions due to the decomposition of the fire-retardant additives. Aluminium trihydroxide exemplifies this type of additive and forms aluminium trioxide with the elimination of three molecules of water in a strongly endothermic reaction. Magnesium hydroxide is similarly used but decomposes at a higher temperature.

Physical action: reduced combustible mass. In general, the coating itself should contain inert fillers at a relatively high phase volume; this will dilute the combustible component during the initial stages of combustion and also act as a heat sink.

Chemical action: radical scavenging. Combustion is a complex chemical reaction where polymer decomposition produces mostly volatile radical species in its initial phase.

These radical easily react with oxygen feeding the combustion.

The presence of additives interfering with such radical reaction represents an opportunity to interrupt or slow down the combustion.

Additives acting as oxygen eliminators give a similar opportunity (Table 9).

3.3.8. Plasticisers

A plasticiser is a chemical compound added to a polymer to make it softer and more flexible by decreasing its glass transition temperature (T_g). Ideally, the plasticiser is 'internal', that is, an integral part of the polymer,

Table 9 Example of fire-retarding additives

Halogen derivates	Phosphorus derivates
Chlorinated paraffins	Phosphoric acid esters (also to aid
Polybromodiphenyl oxides	char formation)
Cl and Br containing polyols (may be used as polyurethane components)	Dibutyldihydroxyethyldiphosphate
Other resins containing halogen	

incorporated during synthesis as a co-monomer or resin. External plasticisers are sometimes used as modifiers. These are usually small molecules with pendant polar and non-polar groups able to both aid solubility, and through steric hindrance prevent inter-molecular interaction of polymer chains thus increasing mobility.

Typical plasticisers include polyesters of adipic, azelaic and sebacic acids with aliphatic diols. It is important that they resist leaching otherwise the film will embrittle, leading to cracking and detachment.

Coalescing agents described earlier are a special case of transient plasticisers used to aid the film formation of dispersion polymers.

4. COLORANTS (PIGMENTS AND DYES)

Colorants are substances added to a coating in order to modify the optical properties of the film, in particular opacity, and the absorption and scattering of light that gives rise to the phenomenon of colour.

Colorants may be opaque or transparent, and the latter may be soluble or insoluble in the coating vehicle. Where soluble, colorants are described as 'dyes' and will penetrate a substrate like wood, possibly becoming insoluble in the process. Pigments, in contrast, remain insoluble throughout the coloration process and must be dispersed rather than dissolved during manufacture of the coating. The colour of pigments and dyes is largely derived from specific chemical groups known as chromophores (from the Latin *chromos* = colour and *fero* = to bring). Historically, many chromophores were first developed as dyes requiring chemical modification to make an insoluble pigmentary form. In some cases, this is achieved by precipitation onto an insoluble inorganic substrate such as alumina or barium sulphate. Such materials are described as 'lakes' whereas pure pigments may be described as 'toners'. The latter nomenclature is more common in the USA; in Europe, the term 'toner' may be used to describe certain specific water-soluble dyestuffs that are co-precipitated with an inorganic compound. Rather confusingly the coatings industry sometimes uses the term 'toner' for a second colour that is added to adjust hue. This process is better described as 'tinting'.

Although pigments can be classified as natural or synthetic, a more relevant generic distinction can be made between inorganic and organic pigments. Inorganic pigments are denser and more opaque and have a lower surface energy, which is significant when it comes to wetting and stabilisation in coating media. Organic pigments are hydrophobic rather than hydrophilic and offer a much wider colour gamut of saturated colours than the inorganic. As complex organic molecules, their synthesis is usually a multi-stage process and like other fine chemicals they are expensive.

The classification of organic pigments is conveniently covered under two headings, either chemistry or colour. Certainly the latter is useful for the user and the well-known 'Colour Index' (http://www.colour-index. org/) groups pigments into colour families. However from the perspective of understanding the origins of different pigment types, it may be more convenient to turn this around into either chemical families or process categories. The former would include division into azo pigments and their many sub-divisions, and polycyclic pigments such as phthalocyanines and perylenes.

Process terminology might include BON (beta-oxy-naphthoic acid) pigments, vat pigments, resinated pigments, flushed pigments as well as the intermediates that are produced at different stages in manufacture and possibly sold separately. It may be noted that while there is a great deal of commonality in classification schemes, different authorities will use different groupings of which the most simple is a distinction between 'azo' and 'non-azo' or polycyclic categories (Table 10).

Although pigments are best known by their interaction with visible light and hence colour, they also may absorb or scatter in other parts of the electromagnetic spectrum. This can give both coatings and coated substrates protection from UV degradation, and for some special pigments reflection of infrared (IR) heat. Pigments may also confer other specific properties including fire retardance and corrosion protection.

4.1. Origin of colorant properties

There are many underlying factors that influence colorant properties including:

• Chemical nature
• Crystal structure

Table 10 Pigments: Chemical families

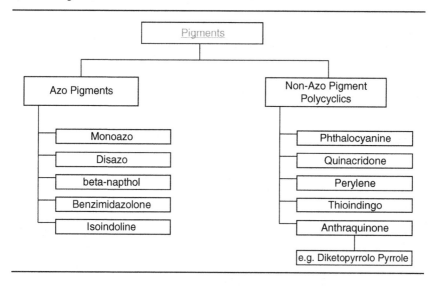

- Particle shape
- Particle size
- Particle size distribution
- Refractive index
- Pigment surface coating

Clearly, the chemical nature has a major effect including the presence of specific chromophores. Chemistry will also have a direct effect on the surface energy and this may manifest itself in properties such as the colloidal stability, and ease of dispersion, in a specific medium. However, it should also be noted that some pigments are coated with other materials to modify surface energy and other properties.

Refractive index is an important intrinsic property of pigments and dyes that affect light scattering. Maximum scattering requires a large refractive index difference between pigment and binder. If the refractive is close the material will appear transparent, this of course is an advantage when the objective is to control the gloss level (sheen) of transparent wood coatings.

The optical properties of pigmented coatings are also strongly influenced by extrinsic physical parameters such as particle size and particle size distribution, hence the importance of the dispersion process. Changes in particle size and its distribution can significantly shift the hue of coloured pigments.

4.2. Required colorant properties

The selection of colorants for a specific application will like all raw materials depend on meeting technical criteria at an economic cost including processing. Key technical factors include:

- Appearance
- Colour
- Tinctorial strength
- Solubility
- Physical form
- Crystal structure
- Particle size
- Opacity/transparency
- Durability

4.2.1. Colour

In principle, it is possible to create a wide colour gamut, from three primary colours through the physical and physiological mechanisms of additive and subtractive colour mixing. The additive primaries are blue, green and red; and the subtractive are cyan magenta and yellow. This is a complex subject [5], and in practical terms there are no pure primary pigments or dyes. Moreover, colour effects are a combination of additive and subtractive with a different balance between opaque and transparent colours. Thus, to achieve a specific hue, such as a deep purple, it may be necessary to use a very specific pigment. However, it is possible to generate large families of colours from relatively few pigment combinations and modern tinting and blending systems illustrate this. Many wood coatings are produced in wood-like colours and the range of colorants required is relatively small.

As noted earlier, the colour of pigments and dyes depends on the amount of light absorbed and reflected. With a few exceptions, the colour of organic molecules derives from absorption by 'chromophores' which are aromatic in nature; the electron donation and acceptance is modified by the presence of 'auxochromes', such as $-NO_2$ or $-COOR$. With inorganic pigments, a different mechanism is involved and charge transfer between molecular orbitals is involved. For a useful description of the origins of colour, see Nassau [6].

4.2.2. Tinctorial strength

Ideally, the coloration of a coating is achieved with a minimum of pigment or dye and this requires high tinctorial strength. The term 'strength' is often used in association with 'pigmented intermediates', and will depend on the chemistry modified by physical factors such as particle size and colloidal stability in the chosen medium. With pigments, the choice of dispersant is particularly important.

4.2.3. Solubility

Solubility and insolubility are defining characteristics of dyes and pigments. A wide range of dyes are employed including reactive and disperse types depend for their action on various solubilisation processes.

Although pigments are described as insoluble, they may show a partial solubility that is generally seen as a defect.

'Blooming' arises when a pigment has some partial solubility in solvent, especially at elevated temperatures, and then re-precipitates as a fine powder not unlike the bloom on some fruit. 'Plating out' is a similar phenomena caused by initial poor wetting.

'Bleeding' occurs if a second coat solubilises part of the first coat. It is often associated with red pigments and dyes, but can occur with other colours.

'Re-crystallisation' can occur under some processing conditions such as high temperature and shear.

4.2.4. Physical form

Pigments can be amorphous but in many cases are crystalline, and can exist in more than one crystalline form (polymorphism) with very different properties.

Colour depends on the ratio of scattering to absorption. Scattering of light rises to a maximum for a given wavelength and then falls again (Fig. 6).

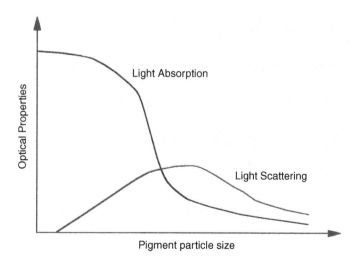

Figure 6 Schematic relationship between pigment size and optical properties.

Thus, titanium dioxide pigment is manufactured to a crystal size typically around 0.24 μm to maximise whiteness and opacity. Nano-size titanium dioxide (80 nm) in contrast is transparent. Dyes and transparent pigments are therefore necessarily in the smaller size range. Coloured pigments that are to have good covering power (opacity) are manufactured to a coarser particle size; however, a finer size will improve tinctorial strength. It is not easy to covert convert larger particle size pigments to finer ones by the normal processing techniques used in the coatings industry. Pigment should always be selected with a view to the intended application.

Most pigments undergo a manufacturing process in which at some stage the pigment will be in the form of wet slurry. The slurry is filtered by squeezing to form 'presscake', which is subsequently dried to a low moisture content. The bulk density of the dried pigment is much lower than the density of the pigment itself. For example, titanium dioxide has a bulk density of ~ 1.0 g cm^{-3}, whereas the density of the pigment is ~ 4.05 g cm^{-3}. Thus, the pigment as supplied contains 75% air, which explains why wetting is an important stage in the dispersion process. Pigments that are to be used in water-borne coatings may be available as strength-controlled presscake, which avoids the wetting out process. Water from presscake may also be displaced by non-aqueous solvents, to produce a 'flushed' dispersion suitable for use in solvent-borne coatings.

4.2.5. Durability

Generally, colorants are required to be durable and resistant to fading or darkening. Wood coatings differ from other market sectors in that coatings may be transparent and the appearance of the coating will also depend on the coated wooden substrate, which will also undergo changes. UV absorbers may help control this. Most pigments will undergo some colour change on exposure to solar radiation though a few inorganic materials are unaffected. Light fastness is often quantified by comparison with a 'Blue Wool Scale' (1 = poor, 8 = excellent), where strips of wool dyed with standard dyes of known fastness are used as a standard of comparison.

Light fastness is important for both internal and external exposure; however, for external conditions the interaction with the binder is also important. Some pigments may promote photo-degradation; for example, titanium dioxide is very reactive unless coated with a coherent coating; oxides of silica, aluminium and zirconium are widely used commercially. Iron oxides give good exterior durability due to their UV and visible light absorption and this is fortuitous for coatings that represent wood shades.

The pigmentation of any binder will also affect other properties including mechanical properties, which will in turn influence durability.

Generally, an increase in the pigment volume concentration of a coating will reduce the extension to break and increase the elastic modulus (stiffness). If taken too far an increase in PVC will also increase a propensity towards cracking and other mechanical failure.

4.2.6. Toxicology
Many pigments have a low solubility and bioavailability; they do not present major environmental hazards. There are exceptions and pigments which have been highly valued in the past such as red and white lead, lead chromate have either been banned or are in the process of being phased out due to unacceptable hazards associated with their use. With modern pigments, health hazards are rigorously assessed and any hazards published through material safety data sheets. Organisations such as the ETAD (Ecological and Toxicological Association of Dyes and Organic Pigments) publish guidelines on safe pigment handling. The ETAD Web site (http://www.etad.com) also provides useful links to pigment manufacturers' Web sites.

4.3. Pigment types

Although pigments can be classified in several different ways, the distinction between inorganic and organic has a number of practical consequences. Major differences are summarised in Table 11.

4.3.1. Inorganic white pigments
White pigments include zinc oxide, zinc sulphide, antimony oxide and basic lead carbonate (white lead).

These retain a few specialist applications, but by far the dominant opacifying pigment is titanium dioxide (TiO_2).

Titanium dioxide is available as anatase and rutile crystalline modifications, the latter being most widely used. Anatase has a lower refractive index than rutile and less absorbency in the short wave region, which is useful for some specific coating applications. However, it finds its main use in the paper and textile delustering industries.

Titanium dioxide pigments are almost invariably coated to improve durability, dispersion, etc. The coating has a marked effect on electrokinetic properties and pH is an important formulating parameter in all water-borne coatings.

The two major methods of production are the sulphate process and the chloride process. The former has higher costs to meet environmental targets and is generally being displaced by the latter for new investments. However, anatase can only be made by the sulphate process.

Table 11 Differences between inorganic and organic pigments

Property	Inorganic pigments	Organic pigments
Colour	Duller shades	Brighter and more saturated
Colouring strength	Normally worse	Can be high
Bleeding	Rare	Varies greatly but possible
Hiding power/ opacity	Usually high	Lower and can be transparent
Refractive index	Depending on composition	Depending on composition
Particle dimension	Coarse – fine	Can be very fine
Chemical reactivity	Depending on composition	Depending on composition
Stability to heat and light	Usually good	Some grades fade
Coating properties	Depending on composition	Depending on composition
Surface energy	High, hydrophilic	Lower, hydrophobic
Cost/price	Can be inexpensive, some high	More expensive, can be very high

4.3.2. Coloured inorganic pigments
Major coloured inorganic pigments are listed below:

- Brown
  - Iron oxide brown
- Yellow
  - Lead chromate
  - Cadmium yellow
  - Yellow oxides
  - Bismuth vanadate
- Red
  - Lead molybdate
  - Cadmium red
  - Red iron oxide
  - Red lead (anti-corrosive function)
- Blue
  - Prussian blue
  - Cobalt blue
  - Ultra-marine (synthetic lapis lazuri)
- Violet
  - Manganese ammonium pyrophosphates

- Green
  - Chrome green
  - Chromium oxide

Of these, the brown, red and yellow iron oxides are particularly useful in wood shades. Iron oxides occur naturally but for the wood coating industry are more like to be synthetic. Synthetic grades can be precipitated and dispersed to give fine particle sizes that are fully transparent.

4.3.3. Coloured organic pigments

For the coloured organic pigments, a schematic sub-division is presented based upon colour and chemistry:

- Organic Reds:
  - Azo pigments metalised and non-metalised (e.g. Naphthol Reds)
  - Nickel complex pigments
  - Diketo pyrrolo pyrrole (DPP)
  - Quinacridone reds
  - Vat reds
  - Dibromanthrone
  - Anthraquinone
  - Brominated pyranthrone
  - Perylene reds
- Organic Greens:
  - Copper phthalocyanine
- Organic Blues:
  - Copper phthalocyanine (alpha and beta)
  - Indanthrone blue
  - Carbazole violet
- Organic Yellows:
  - Monoarylides
  - Diarylides
  - Benzimidazoles (azo derived)
  - Heterocyclics
- Organic Oranges:
  - Azo based
  - Benzimidazole

A detailed account of the chemistry and properties of these pigments is outside the scope of this book; for a useful overview, see Lewis [7].

4.3.4. Extender pigments

Extenders are inorganic substances similar to pigments, they may be known also as 'fillers' or 'inerts'. Very often, the distinction between pigments and extenders is not sharp being based partly on economic criteria. However, a major distinction is that they have a relatively low refractive index, comparable to that of the binder, and can therefore confer relatively little opacity. The definition of extender is given by EN 971 as *a material in granular or powder form, practically insoluble in the application medium and used as a constituent of paints to modify or influence certain physical properties.*

Extenders can be divided into four main chemical groups (Table 12) and can be natural or synthetic. They vary widely in particle size and shape; the coarser grades must be milled or ground into the surface coatings. The harder types such as barium sulphate resist breakdown by normal processes and should be purchased in a micronised form.

Crystalline natural extenders have three general shapes: modular, acicular and lamellar. The modular varieties tend to settle compactly in paints. The acicular generally settle to easily stirred conditions and can reinforce dried films. They tend to lower the gloss and improve topcoat adhesion by promoting surface irregularity. Lamellar extenders such as mica also reinforce films and, by overlapping in leaf or plate fashion, help to improve resistance to water transmission.

Extenders are used in different ways to modify coating behaviour including:

- Cost reduction (particularly in decorative matt coatings)
- Rheology and viscosity
- Film reinforcement
- Filling and sanding properties
- Reflectance (gloss and sheen)

Table 12 Extender pigments

Sulphates	Carbonates	Silicates (kaolinite, smectites, amphibolites, zeolites)	Oxides
Barium sulphate (Barytes)	Calcium carbonate (Chalk, Whiting)	Aluminium silicate natural (China Clay) (Bentonite)	Silicon oxide synthetic (Silica)
Calcium sulphate (Gypsum)	Magnesium carbonate (Magnesite)	Aluminium silicate synthetic	Silicon oxide natural (Quartz)
Strontium sulphate (Celestine)	Barium carbonate (Whiterite)	Magnesium silicate (Talc)	Aluminium oxide (Alumina)
		Potassium aluminium silicate (Mica)	Magnesium oxide

The last function, already mentioned under 'matting agents', is particularly important for wood varnishes and lacquers. Silicates are widely used including:

- Precipitated silicas and silicates
- Silica gels, sols, colloids
- Silicon tetrachloride derived (fumed:pyrogenic)

REFERENCES

[1] Ho, D. L., and Glinka, C. J. (2004). New insights into Hansen's solubility parameters. *J. Polym. Sci. Polym. Phys.* **42**(23), 4337–4343.
[2] Stewart, P. A. (1995). Literature review of polymer latex film formation and particle coalescence (http://www.initium.demon.co.uk/filmform.htm).
[3] Bieleman, J., ed. (2000). "Additives for Coatings", 372 pp. Wiley-VCH Verlag, Weinheim (ISBN 3-527-29785-5).
[4] Marrion, A. R. (1994). "Chemistry and Physics of Coatings", 206 pp. Royal Society of Chemistry, Cambridge (ISBN 0-85186-994-7).
[5] Berns, R. S. (2000). "Billmeyer and Saltzman's Principles of Colour Technology", 3rd edn, 247 pp. John Wiley & Sons, New York (ISBN 0-471-19459-X).
[6] Nassau, K. (1987). The fifteen causes of colour: The physics and chemistry of colour. *Color Res. Appl.* **12**(1), 4–26.
[7] Lewis, P. A. (2000). "Organic Pigments", 3rd edn, 43 pp. Federation Series on Coatings Technology. Federation of Societies for Coatings Technology, Blue Bell, PA.

BIBLIOGRAPHY

Bulian, F. (2008). "Verniciare il legno." Hoepli, Milano.
Di Battista, P. (1994). Introduzione teorico-scientifica ai principi della fotopolimerizzazione, Atti seminario Radtech/CATAS, Udine.
Mellan, I. (1977). Industrial Solvents Handbook, Noyes Data Corporation.
Van Esch, G. J. (1997). Flame Retardants: A General Introduction, Environmental Health Criteria 192, World Health Organization, Geneva.

CHAPTER 5

Classification and Formulation of Wood Coatings

1. INTRODUCTION

Many useful compositions are defined by a list of ingredients, also described as a 'recipe' or 'formula', upon which a series of operations must be carried out to deliver the required user benefits. The term 'formulation' is thus both a description and a process. Examples can be found in industries ranging from food, rubber tyres, pharmaceuticals, polishes, sun creams and toiletries to adhesives, ink and coatings. The needs of a specific industry are met by making an informed selection of components (i.e. the Raw Materials described in Chapters 2 and 3), combining them in specific ratios and in the correct sequence at a specified state of sub-division. Clearly, the 'formula' itself is a key piece of intellectual property or 'know-how' which is integral to the whole of the supply chain from procurement to delivery.

The task of the formulator is to meet business needs by designing a product which has the right appearance, and a balance of properties suited

Wood Coatings: Theory and Practice
DOI: 10.1016/B978-0-444-52840-7.00005-9

to the envisaged application. This must include operational requirements as well as physical properties. Operational requirements are different between decorative and industrial markets but will include aspects such as application and storage, and are discussed in more detail in Chapters 7 and 8. The majority of decorative products are supplied ready for use (RFU), but for industrial markets products may need thinning or activating (two-pack products or 2K) and may have limited pot lives.

It might be assumed that all wood coating would have something in common to distinguish them from other categories of coating, for example, for metal or masonry. While this is broadly true, it is not usually apparent from the formulation alone. Consider, for example, the difference between an impregnating stain for decking, and the finish used on a white kitchen cabinet. There will be no common ingredients even though both have been formulated to meet the overarching needs of wood (or wood derived) substrates as described in Chapter 2. It is important to recognise that formulating a wood coating imposes certain constraints on the selection of ingredients, but that there are still many choices and degrees of freedom on the permutations and combinations of raw materials that can be used to meet a defined need.

Formulation has been defined as 'The science and technology of producing a physical mixture of two or more components, (ingredients) with more than one conflicting measure of product quality' [1]. This definition emphasises two important aspects of formulation. Firstly that the emphasis is on the 'physical mixture' properties; chemistry is involved in making ingredients but the physical properties of the wet and dry film are dominated by mixture properties, in particular the ratios between components. A second aspect of the definition underlines that there are conflicting measure of quality. This partly explains why there are so many different types of wood coating; they present alternative solutions to the formulating problem. However, even within closely related product groups compromises will be necessary. For example, the rheology that prevents sagging may not be ideal for flow and levelling. Another overarching consideration is that of cost and the contribution of expensive ingredients must be fully optimised.

There are many perspectives from which formulation can be considered; for the purposes of this chapter, three broad areas are considered:

(1) Classification
(2) Mixture properties
(3) Composition

The 'classification' of wood coatings by various criteria is almost equivalent to a specification. It defines the type of product that is required and is the first stage of selecting formulation options.

Mixture properties are at the heart of formulation. All the components exist in specific relationships which will be reflected in the formula. Some components relate to the whole of the wet or dry film in order to achieve a concentration effect (e.g. the amount of wet or dry film biocide), others such as catalysts or curing agents will relate to specific elements of the binder. For many properties, the dominant influence is the ratio of pigment to binder and the total solids content.

Composition covers the nature of the main film former (the binder) and the technology by which it is to be delivered. It also covers pigmentation and any additives necessary to make the formulation viable.

In this chapter, classification and mixture properties are described; compositional aspects are taken up in Chapters 7 and 8.

2. CLASSIFICATION

Classification is a powerful tool in science and has imposed order into diverse fields such as medicine, chemistry and biology. Classification is often based groupings showing similarity, that is, something in common. However as noted above, many wood coatings (a classification term based on the substrate) do not have much in common, save for the most general of properties which would also apply to other coating types. It is therefore to consider sub-categories of classification, some of which will specific to wood and others which may be more generic. From such a framework, it is then possible to define a wood coating type in terms that enable a formulation specification to be drawn up.

An overview of some potential classification categories is given in Table 1.

Table 1 Wood coatings. Potential classification categories

Classification term[a]	Example
Generic type	Paint, varnish, stain
System function	Primer, basecoat, finish
Solids content	Low solids, high solids
Film build	Non-filming (e.g. impregnating stains, waxes); film forming (paints, varnishes)
Appearance	Colour, gloss, build
Chemistry	Alkyd, acrylic, polyurethane
Technology	Water-borne, solvent-borne, UV curing (Radcure)
Delivery	Air drying, stoving, two-pack
Property	Flexible, permeable, fungal resistant
Market sector	DIY, industrial joinery, interior, exterior
End use	Flooring, joinery, furniture

[a] It would be unusual to use all of these categories in a descriptive sense, for example, 'a semi-transparent red glossy air drying water-borne flexible acrylic basecoat for garden furniture'! In most situations, only part of the information is needed and some is assumed. Nonetheless to create such a product, it would be necessary to incorporate the terms into an objective and consider the implication for formulation.

2.1. Generic type

2.1.1. Paint

Paint is a very broad generic type and was formally defined in the introduction to Chapter 3. It is generally assumed that paints are relatively opaque. Paints are used on all types of substrate and to be suitable for wood must have specific properties relating to flexibility, adhesion and permeability according to the end use. Interior and exterior requirements will be significantly different; this aspect is discussed further in the following chapters. The term 'paint' may also incorporate other system functions such as priming (see below), and it is thus possible to describe a wood primer, or undercoat as a paint and the subsequent layers as 'topcoat' or 'finish'.

2.1.2. Clear and semi-transparent coatings (varnishes and lacquers)

One of the unique features of wood as a substrate is the attractive figure, and there is thus a strong requirement for clear (i.e. transparent) coatings to enhance the appearance of wood. Transparency does not preclude colour though it does limit the agents (pigments, dyes, stains) that can be used in comparison to paint, and the colour of the substrate itself will also influence the final effect.

The classification terminology for clear and semi-transparent wood coatings is undergoing change, and may be used differently by users in different sectors. Strictly speaking, a 'varnish' should dry exclusively by oxidation though the term may be used for other technologies. Varnish is translated as 'Vernici' or 'Vernis' in Italian and French, but there is no equivalent German term where 'Klarlack' is used. The term clearcoat or clear coating (or its equivalent) is also common in other languages. The term 'lacquer' was defined as a transparent material that dries solely by evaporation of solvents, but it was accepted that stoving lacquers did dry by other means (cross-linking), and the term is sometimes used more generically to denote a clear coating. In English the term 'glaze coat' was used to describe a clear layer used over some kind of ground coat. This terminology has developed in the car industry where 'Clear over Base' or 'Base Coat Clear' systems are common.

2.1.3. Stains and lasures

The classification terminology for stains is also complex and evolving. Traditional stains were either non-film forming, or penetrated the wood sufficiently that no film could be measured. Stains can contain dyes or sometimes chemicals that react with wood including acids such as oxalic acid, bleaches and caustic alkalis.

A major use of stains is to enhance, or 'ennoble', the characteristic appearance of wood prior to coating with a transparent finish. Thus in the UK, stains were defined as materials to colour a substrate by penetration without hiding it. Stains were further sub-divided according to the vehicle used and might be classed as water stains, oil stains or spirit stains. The word 'impregnating' (Latin *in-pregno*), is often associated with stains, as it is differential penetration which enhances wood grain. Impregnation of the substrate is also useful for biocidal protection, and to enhance adhesion to the substrate.

The terminology of stains became more complex when they were used externally as an alternative to paint systems. The term 'lasure' is widely used in Europe to describe such products although in the UK either 'exterior stain' or a proprietary term is more common. Lasures will contain pigments rather than dyes, and in many cases biocidal agents to inhibit blue stain in service. For reasons discussed in the next chapter, the formulation of stains has evolved to increase their solids content; they become more film formers than impregnating and may be described, for example, as 'high solids exterior stains'. The term high solids is relative, since the solids content is still low compared to that of most paints. A further complication arises from the fact that some manufacturers adopting the formulating principles of exterior stains but with opaque pigments. The term 'opaque stain' was thus coined, though it is regarded as a contradiction by some. It was against this background that the European Standards Committee for exterior wood coatings (CEN TC139/WG2) developed a broader classification system to cover build (film thickness), transparency and gloss level.

EN 927-1 defines four levels of build, three levels of transparency and five levels of gloss, enabling 60 different products to be described. The classification provides a description for currently available coatings and has the provision to describe new categories. The following list illustrates how some typical coating systems might be classified using these criteria. The terms are descriptions and not precise definitions (Table 2).

In the USA, semi-transparent coatings for wood may be described as 'water-repellent preservative stains' (WRP or WRPS). Their origin may be

Table 2 Classification of typical coating systems according to EN 927-1

	Build	Transparency	Gloss
Alkyd gloss paint system	High build	Opaque	High gloss
Latex gloss system	Medium build	Opaque	Gloss
Alkyd varnish (three coats)	High build	Transparent	High gloss
Joinery wood stain	Medium build	Semi-transparent	Semi-gloss
Fence surface treatment	Minimal build	Semi-transparent	Matt

traced to the 'Madison Formula' which was developed by the US Forest Products Laboratory in 1950. It contained oil, wax and a pentachlorophenol fungicide. The DIY version was withdrawn from retail sale in 1985 (due to the penta content) but various derivatives still exist [2].

2.1.4. Oils, polishes and patinas

Treatment of wood with oils and waxes is a traditional approach which enhances the natural wood grain and appearance. Examples of exterior treatments include the original 'Madison Formula' (see above), and products that may be described as Teak Oil and Danish Wood Oil (see next chapter). Interior products also include polishes which may also be coloured. Waxes, used for products in interior environments, can be either semi-solid or liquid depending on the amount of solvent used (usually hydrocarbon mixtures). Sometimes they are based upon emulsions of wax in water. They may be described as 'pastes', 'creams' or 'liquid dispersions' according to their consistency.

Waxes are often coloured with iron oxide (red, brown and yellow shades) pigments. They can be applied onto bare wood or after the application of a priming coat. Because waxes do not constitute a real solid film, the protective effect is rather limited and frequent maintenance will be required, especially for those products subject to intense deterioration (floorings). Although not classed as film formers, the repeated use of wax, in combination with light and in-use factors, gradually builds up a characteristically coloured 'layer' on wood which is sometimes described as a 'patina'. The use of this term is denigrated by some on the grounds that a true patina is a deliberately applied coating or treatment more associated with metals such as copper. Nonetheless, the term 'patina' is sometime used in the context of antique furniture and musical instruments like a violin back. A few companies have developed products which are described as patinas. Such products are usually a two-stage wood finish consisting of a 'Grain Enhancer and Sealer' and a 'Final Finish'.

2.2. Functional classifications

The majority of wood coatings, both industrial and decorative, may be considered as a system comprising more than one product. Individual products within a system are described by additional names which indicate their functionality, such as 'primer', 'undercoat', 'basecoat', topcoat, etc. This systems approach recognises that meeting all the characteristics necessary to achieve optimum protective and decorative functions in one product would entail a compromise. A better approach is often to separate the functions so that there is a broad separation between coating/substrate and coating/service environment interactions. A clear-cut distinction can thus be made between primers and topcoats. The function of a primer will depend on the substrate and would clearly differ between say wood, metal and masonry. In the case of wood, the role of the primer includes adhesion promotion and control of penetration.

The nature of the priming coat will depend upon the type of finish to be applied; thus, primers can range from un-pigmented solutions to pigmented paints with specified properties. Sometimes the term 'primer' is replaced by an alternative descriptive term such as 'penetrating base-coat'. Nonetheless, such products still have a priming function and will be over-coated with a finishing coat. It would not be true to say that *all* wood coatings employ a primer; for example, some medium build stains are effectively self-priming, with two or more coats of the same product often being applied. However, a majority of higher build systems in both decorative and industrial sectors do involve separate priming coats.

Between the primer and the topcoat there may be other layers which contribute to protective properties, film build, opacity and colour. Descriptive terms include undercoat and basecoat. In the furniture industry, additional products may be used according to the finish that is desired. 'Sizes' and 'washcoats' are employed to raise and stiffen wood grain, prior to sanding, and also assist in evening out differential absorption of staining coats. For higher build systems, 'fillers' are used both to colour wood pores and to fill them where a high build rather than open-pore finish is required. Having applied filler, further 'sealers' may become necessary to correct differential surface absorption and provide a basis for uniform top-coating. Zinc stearate may be incorporated to aid sanding. The subsequent topcoat can be based on the technologies described in Chapter 3 according to market sector requirements. This aspect of formulation is taken up in Chapters 7 and 8.

3. MIXTURE PROPERTIES

3.1. Introduction

Formulation as a process involves many stages in order to bring a product into full production. At the heart of the process is the need to compile the components (i.e. the raw materials) that will be used to make up the formulation. It has already been shown that these can be grouped into four major categories:

(1) Binders
(2) Solvents (or other volatile materials)
(3) Pigments
(4) Additives

Specific choices within these broad categories must be decided on the basis of a set of business objectives within the context of a market sector end-use category.

Normally, the first stage in formulating is to select the binder technology. This means the chemistry of the binder (e.g. acrylic, urethane, etc.) and the means by which it will be delivered (water-borne, solvent-borne, Radcure, etc.). This choice effectively fixes the solvent options and will have other implications for conversion to a dry film, that is, will the product be air drying or requiring stoving? Will it be one-pack ready mixed, or two-pack?

Having selected an appropriate technology, the next step (except for clear coatings) is pigmentation, the choice of which is governed by considerations such as colour, opacity, transparency and protective needs.

These stages define what might be called a 'framework formulation'; the next step is to introduce the additives (Chapter 4) to make the formulation viable (e.g. pigment dispersants), and other additives to confer additional properties. In general, the choice of additives will be specific to the chosen technology.

This stage of formulation usually requires considerable laboratory experimentation and result in the familiar list of ingredients known as a 'formula' or 'recipe' (Table 3).

Table 3 Recipe of a high solids semi-transparent alkyd wood stain (*source*: EN 927-3:2000 the 'ICP')

Raw material	Trade Name®	Supplier	Active ingredient	% Weight
Long oil alkyd	Synolac 6005W	Cray Valley	65% NV	52.82
Pigment red	Sicoflush L2817	BASF	40% pigment	4.63
Pigment yellow	Sicoflush L1916	BASF	40% pigment	2.30
Rheological additive	Bentone 34	Rheox	10% premix	0.60
Calcium drier	Nuodex Ca	Servo	5% metal 55% NV	2.77
Cobalt drier	Nuodex Co	Servo	10% metal 75% NV	0.37
Zirconium drier	Nuodex Zr	Servo	12% metal 45% NV	0.30
Biocide	Preventol A5	Bayer	90%	0.72
HALS	Tinuvin 292	Ciba-Geigy	100%	0.45
Anti-skin	Exkin 2		100% MEK	0.20
Solvent	Varsol 40	Exxon		34.84

It is common to express a formula in terms of the % weight of the components. To make a litre of the product, the % weight is multiplied by $10 \times$ s.g., where s.g. is the specific gravity of the composition. Formulators will have access to spreadsheets or dedicated software with a database of raw material constants which readily enable the calculation of s.g. and other paint constants.

Although the formulation is expressed as 100% by weight, it embodies a number of key ratios which relate the ingredients to each other in functional terms. Thus in the above example, the % of catalytic driers is based on a relationship between metal content and the auto-oxidisable component of the alkyd. Biocide level is set to achieve a necessary concentration in the dried film. Each component is at a concentration that should be optimised in relation to the functionality required. Should the formulation be changed then the relationship of all the components will need to be re-balanced and all the % weight figures may show relative change.

The formula illustrated above could be described as a 'solvent-borne high solids semi-transparent wood stain', and is alkyd based. In principle, it would be possible to make an equivalent water-borne based on an alkyd emulsion or a water-borne acrylic dispersion. Clearly, such changes would affect both the wet and dry properties of the product but all could still be described as 'wood stains'. The important question thus arises what generic aspects of the formulation define a 'wood stain'. The answer to this question lies primarily in two important ratios embedded in the formulation, namely the 'pigment:binder' ratio, and the 'total solids content'. This aspect of the mixture properties is discussed in the next section.

3.2. Pigment-to-binder ratios

The amount of non-volatile material in a paint is usually termed 'the solids content', expressed in either weight or volume terms. 'Volume solids' is most relevant in terms of the contribution to film properties; 'weight solids' is usually preferred for specification and recipe (formulation) expression. The two options are inter-convertible through the relationship: volume = mass/density. The area over which the coating is spread and the volume solids will decide the film thickness after drying (Fig. 5).

The bar chart (Fig. 1) represents four simple model coatings each having a volume solids content of 50%. Coatings 2–4 have an increasing ratio of pigment-to-binder volume.

For a given set of ingredients (raw materials), the volume ratio of pigment to binder (the P/V ratio) *in the dry film* has a dominant effect on film properties.

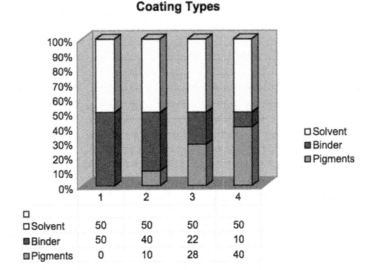

Figure 1   Schematic illustration of *P:B* ratios.

Another way of expressing the P/V ratio is either as a decimal fractional concentration of pigment in the total film volume, or as a percentage – known as the 'pigment volume concentration' (PVC):

$$PVC = \frac{V_{pig}}{V_{pig} + V_{bind}} \quad \text{or} \quad PVC\% = \frac{V_{pig} \times 100}{V_{pig} + V_{bind}}.$$

Alternatively, the pigment-to-binder relationship may be expressed in weight terms expressed as the P/B ratio. Clearly P/V and P/B are inter-convertible if the density (or specific gravity) of the components is known. As ratios P/V and P/B can be in any numerical form, but it is common to set the binder to unity, thus a ratio of say 27:84 would be reported as 0.32:1 or just 0.32.

P/B, P/V, PVC CALCULATION AND CONVERSION

- PVC = pigment volume/(pigment volume + binder volume)
- P/B weight = $D/(1/PVC) - 1 = PV \times D$
- P/V volume = $1/(1/PVC) - 1 = (P/B)/D$
- $PVC = 1/\{(D/P:B)+1\}$
- $PVC = 1/\{(1/P:V)+1\}$

where D is the pigment:binder density ratio, P/B is the pigment:binder ratio by weight and P/V is the pigment:binder ratio by volume.

When a coating has a relatively low PVC, the pigment particles will be well spread out and the film will be glossy. As the PVC increases, the particles (or pigment clusters) come closer together. Eventually, a point is reached where they touch and there is just enough binder to fill the voids between the pigment particles, this is known as the 'critical PVC' (CPVC). At even higher pigment contents, there will be insufficient binder to fill the voids (after the solvent or water evaporates) and the dry film will become porous. The properties of the four coatings illustrated above can thus be summarised as presented in Table 4.

As noted above, the PVC at which there is no longer enough binder to fill the spaces between the pigment particles after evaporation of all the liquid is known as the critical CPVC. Paints formulated below or above the PVC will have very different properties, since the latter will contain air voids (pores). The porosity can be estimated from the simple expression:

$$\text{Overall porosity} = 1 - \frac{\text{CPVC}}{\text{PVC}}.$$

Thus in the case of coating 4, if the CPVC is 0.6, then the porosity will be $1 - (0.6/0.8) = 0.25$ or 25%.

The value of the CPVC is dependent on factors such as pigment packing, particle size distribution and degree of dispersion. In principle, it can be calculated but is more likely to be determined experimentally. An estimate of the CPVC can be made from the oil absorption of the pigment. Oil absorption OA is the amount in grams of linseed oil of a specified acid value expressed as weight in grams of oil required to wet

Table 4 Summary of the properties of the four coatings described in Fig. 1

Coating reference	General nature	Volume solids (%)	Weight solids (%)	PVC	P/B ratio:1
1	Glossy clear film	50	51.22	0	0
2	Glossy opaque film	50	62.26	0.20	0.96
3	Low gloss opaque film	50	73.19	0.56	4.9
4	Matt porous opaque film	50	77.53	0.80	15.42

Note 1. For illustrative purposes, the densities of solvent, binder and pigment have been taken as 1.00, 1.05 and 4.05, respectively.
Note 2. Many formulators prefer to express PVC as a percentage rather than a decimal, thus 20% rather than 0.2; it is the decimal value that must be used in the conversion factors given above.

out 100 g of the pigment under test. The test conditions and end point are carefully defined. Since it is defined by weight it is numerically opposite to CPVC, that is, to say a low oil absorption means a relatively high CPVC. Although OA is useful, and widely quoted it does not take into account the effect on packing that different dispersants and coating binders will have. For linseed oil, binder CPVC may be calculated using the expression:

$$CPVC = 1/[1 + (OA \times \rho/93.5)],$$

where ρ is the pigment density.

Interior paints formulated above the CPVC are very common in the decorative architectural sector as matt paints used on ceilings and walls. The presence of voids greatly enhances the scattering power of titanium dioxide, a phenomenon known as 'dry hiding'. However, because the films are totally porous they have poor scrub and stain resistance. It is most *unlikely* to use above-critical formulations on wood coatings; however, this does not mean that the concept of CPVC is not important. CPVC is relevant to all pigmented coatings, regardless of the PVC at which they formulated. This is because the CPVC gives information about the state of dispersion and packing in the film. Mechanical properties of pigmented coatings depend to some extent upon the amount of 'free binder' that is left after the surfaces of the pigment, and the requirements of pigment packing geometry, are met. Hence, models for predicting properties such as rheology will contain a normalising function $(1 - PVC/CPVC)$ [3]. It is also for this reason that the term 'reduced PVC' (Λ), defined as $\Lambda = PVC/CPVC$ is sometimes used as a basis for formula substitution [4]. For example, when substituting alternative raw materials into an established formulation, the levels of pigmentation are adjusted to maintain a constant value of Λ.

3.3. Graphical representation of a coating formulation

Components of a formation, as already noted, are conventionally represented in the form of a list a components, or recipe. A spreadsheet is particularly useful for this purpose and enables paint constants and scaling factors to be easily calculated. The spreadsheet for also be used to represent key ratios in a graphical form such as the bar charts shown above. There is another way to represent key components which is convenient for both visualisation and as a basis for the experimental design of mixtures. Mixture diagrams using tri-linear coordinates are familiar in the form of phase diagrams; they are also convenient for representing a formulation framework.

If a paint is described as a three-component mixture (PBS), it meets the constraints that $P + B + S = 1$ (or alternatively $= 100\%$). PB&W are control

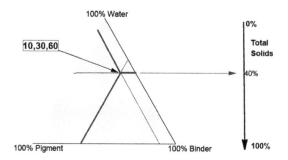

Figure 2   Pigment:binder:water as tri-linear coordinates.

variables but only two of them can be prescribed independently. This means that they can be represented in a (3 − 1 = 2) two-dimensional region termed a 'simplex'. The paint illustrated has an approximate volume composition of 10% pigment, 30% binder and 60% 'solvent'; thus, the percentage coordinates are [10, 30, 60], only two of which have to be specified in order to plot the position as shown below.

A plot such as shown in Fig. 2 has three sets of coordinates, each running parallel to the sides of the equilateral triangle on a 0–100% scale. As plotted here it can be seen that a solids content scale can also be drawn alongside the solvent axis. The volume solids content of 40% is thus represented and is equivalent to 60% on the solvent axis. Mixture diagrams such as this have many useful properties, for example, if two different mixtures are plotted on the same graph, and a line drawn between them, then this represents all the compositions that can be obtained by blending – a practice often described as masterbatching. Another useful element is that the mixture diagram represents a compositional space onto which response surfaces or contours can be plotted.

As an alternative to the three intersecting composition coordinates, it is possible to draw three lines from the apexes of the triangle, through the composition to the opposite side. These are lines of constant ratio composition. Thus, the line drawn from the water apex to the P–B side is a line of constant P:B ratio. The drying paint follows this line until at the P–B-axis (zero water) it represents the PVC of the coating (Fig. 3).

The line that bisects the liquid:binder axis shows the composition of binder in the total medium (in solvent-based paints, it is often called the vehicle solids). The line bisecting the pigment: water axis has no specific meaning in this case but could be related to the composition of the millbase. There are thus six coordinates which are associated with any composition and may have some direct relevance to formulating. However, the caveat remains that there are still only two independent variables, and

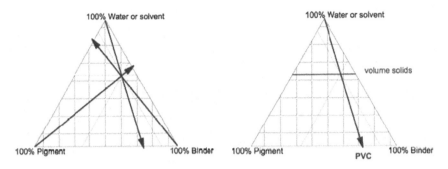

Figure 3 Tri-linear diagrams – additional coordinates.

hence only two of the six need to be specified for the others to be implicit. The two most important coordinates for indicating the essential character of a coating are the PVC and the volume solids.

If these two are specified, then the formulation of some typical wood coatings is shown in Fig. 4. Thus, varnishes will lie on the solvent (including water) binder axis and be relatively high solids. Penetrating stains are in the low solids area and contain very little pigment. High solids stains are intermediate between the two latter and this accounts for certain property differences such as penetration, film build and weathering performance. Moving from stains towards the pigment apex goes through the region that is sometime called 'opaque stains', which is relatively low solids opaque products that might be used for garden furniture, fencing, etc. Many of these will be known only by proprietary rather than generic names. Moving further in the pigment direction is the domain of paints, where primers finishes and topcoats form another cluster.

It should be emphasised that this diagram is not meant to imply that all coating types can be made from the same starting materials. Other

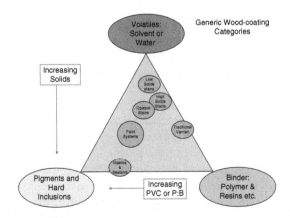

Figure 4 Graphical representation of some typical wood coatings.

factors will dictate the choice of pigment and binder for specific products. Nonetheless, the position within the formulating diagram indicates broad product characteristics.

All coating properties change in a systematic, but non-linear manner, if the PVC is changed [5]. This is common practice for formulators to use PVC as a basis of substation for alternative raw materials, and to make small adjustments to PVC when optimising a formula. Likewise adjustment to the volume solids concentration can be used to change spreading rates and dry film thickness.

It can be seen from Fig. 5 that a combination of the coating volume solids, and the spreading rate, will have a direct influence on film thickness. The theoretical wet film thickness is obtained by the simple expression:

$$\text{Wet film thickness } (\mu m) = 1000/\text{spreading rate m}^2 \text{ l}^{-1}.$$

The theoretical dry film thickness may be obtained from the fractional volume solids of the paint. Thus, a litre of paint spread over 10 m^2, and a solids content of 50% will have a wet film thickness of 100 μm and a dry film thickness of 100 μm × 0.5 = 50 μm.

Clearly when comparing different types of coating, for example, paints versus stain, the dry film thickness will have a major influence on performance in moderating the intrinsic properties of the coating.

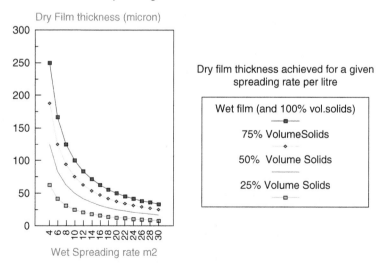

Figure 5 Spreading rate and film thickness.

3.4. Formulation protocols

There is no single strategy for product formulation. Actions taken will depend on the market sector and the starting position. One possibility is to start from an existing formulation and make relatively minor modifications. Starting formulations will be available in the formula databases of major manufacturers, and can be found in patents and other published sources (Ref. 6).

The latter are often derived from raw material manufacturers and their publications which are increasingly accessible from Internet sources. In some circumstances, it may be necessary to start in a 'green field' situation and create a new formulation, though this is unlikely to be free of operational and other constraints. Clearly, the criteria used at each stage will depend on the market sector. Whatever the situation it may be expected that most of the elements illustrated in the figure 6 below will need to be taken into account:

(1) *Agree specification*. This is a vital step; formulation cannot proceed without a clear definition of the target market and any constraints that may be operating including costs, environmental and operational factors. The specification should address key wet and dry physical properties taking into account the substrate and target markets.

(2) *Understand the 'property weightings'*. This is an extension of the first step. It means putting some value on properties which conflict to arrive at the best compromise. Formulation always involves compromises! This stage ideally involves customer consultation. 'Value analysis' is a useful technique.

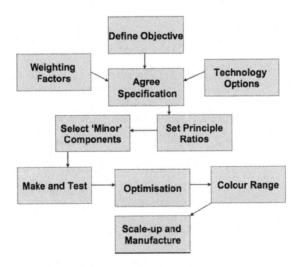

Figure 6   Stages in formulation.

(3) *Select a starting point.* For a raw material supplier, this will usually mean building products around specific technologies or materials. For the paint company, this could be a green field situation and implies selecting or synthesising the right binder technology. Thus, the formulator must start with an appreciation of the practical implications and limitations of the chosen generic technology. Does the technology have the capability to deliver the right property mix? In practice, the choice of technology may also be constrained by operational factors such as the layout and capacity of existing plant. Increasingly, the start point is governed by legislative aspects in relation to compositional elements such as VOC.

(4) *Set ratios of principal components.* This is a fundamental part of generic formulating. As discussed above, the most important ratios are that of pigment:binder (PVC) and volatile:non-volatile (volume solids) which define the essential character of a formulation.

(5) *Select minor components.* These will include so-called 'additives' or 'modifiers', including pigment dispersants, rheology modifiers, coalescing aids, biocides, driers, etc. In each case, the basis of selection and level set will depend on the technology and should reflect the mechanistic function. Thus, coalescents and driers will relate to binder, rather than total volume content, and pigment dispersants should have some relationship to pigment surface area but modified by rheological interactions.

(6) *Test and re-test.* Some knowledge of repeatability and test capability is necessary to make informed decisions. For certain products ascertaining storage stability, or durability, might become rate-determining factors and should be initiated at the earliest possible stage. Application as well as dry film properties must be checked and for industrial finishes this will involve line trials.

(7) *Fine tune and optimise.* This is where the final compromises must be made to maximise the value of desirable properties within operational and financial targets. Experimental design techniques are usefully deployed here.

(8) *Colours and tinting.* Most paint systems need to be available in a range of colours, for decorative and re-finish paints this will be very extensive. Formulating protocols will often extensively develop a single product and then derive colours on a basis of substitution, which will require a re-balancing of all the components. There are important operational differences between colours that are made by a 'co-grind' route, or by blending or tinting. The latter may be in the factory or at the point of sale.

(9) *Scale up and manufacture.* This too is a major step in formulation. Scale up is based on a premise of similarities but it will not be possible to keep all geometric, kinetic and dynamic ratios constant, hence

differences in properties can arise on scale up that will require changes to the process if laboratory- and plant-made products are to align. An important aspect of manufacture is the setting up of control procedures that relate control limits to the product specification in a meaningful way.

Formulation aspects of specific product types are discussed in the next two chapters.

3.5. Relevance of formulation data to users

Much of the information discussed above and in preceding chapters is relevant to creating a formulation. In the cases of users or intermediaries, such as Architects, Specifiers, Purchasing Departments, etc., more emphasis is placed on accurate product descriptions. Such information is readily available through product literature including the material safety data-sheet (MSDS). Product information should provide a clear guide to suitability for a particular end use including operational and environmental information. The classification and selection guide for exterior wood coatings (EN 927-1; Table 3) contains a table to aid this process which may be available from some manufacturers.

Other information such as the nature of the binder may be available through the MSDS and can provide some expectation on properties in relation to the chosen chemistry and technology. The PVC of a coating is seldom disclosed directly but may be determined by analysis. Volume (or weight) solids are more likely to be disclosed and can also be readily determined by analysis. The 'volume solids' gives a useful indication of average film thickness expected from the application spreading rate (Fig. 5).

REFERENCES

[1] Bohl, A. H. (1990). "Computer Aided Formulation." VCH Publishers, New York.
[2] Williams, R. S. (2002). "Alternatives to the Madison Formula." US Forest Products Laboratory, Madison, WI (http://www.fpl.fs.fed.us/documnts/finlines/willi02a.pdf).
[3] Krieger, I. M., and Dougherty, T. J. (1959). A mechanism for non-Newtonian flow in suspensions of rigid spheres. *Trans. Soc. Rheol.* **3**, 137–152.
[4] Bierwagen, G. P., and Hay, T. K. (1975). The reduced pigment volume concentration as an important parameter in interpreting and predicting the properties of organic coatings. *Prog. Org. Coatings* **3**, 281–303.
[5] Patton, T. C. (1979). "Paint Flow and Pigment Dispersion", Fig. 7.1. John Wiley & Sons, New York.
[6] D'Antonio, C. (1998). "Paint Formulations with low VOC." Publisher's Cada, Salerno (Italy).

CHAPTER 6

Properties of Wood Coatings – Testing and Characterisation

Contents			

1. INTRODUCTION

The finishing of wood surfaces with paints and varnishes is carried out with the objectives of achieving the desired appearance, maintaining this appearance during service and in many cases to confer protection to the

Wood Coatings: Theory and Practice
DOI: 10.1016/B978-0-444-52840-7.00006-0

final product. Overall performance depends on several factors, in particular the substrate, the coating system and the interactions between them. Performance also depends on operational factors such as preparation and application, and consistency in production, in other words good quality control.

In the words of Lord Kelvin, 'If you cannot measure it you cannot improve it'. Testing is at the heart of both scientific and technological disciplines. Test methods have many different objectives; they are used to verify the quality of raw materials and to monitor production processes from manufacture to application. Moreover, tests are used in product development to quantify performance characteristics in service, or in a predictive manner by evaluating the influence of fundamental properties on performance. In this case, it is important that the correlation between the property and the performance criteria is well understood; unfortunately, this is not always the case. In consequence, there is often a choice of strategies for testing. For example, the exterior durability of wood coatings might be quantified by long-term exposure testing under conditions of natural weathering under controlled conditions, or by statistical examination of real buildings. This is time consuming, particularly at the development phase and alternative strategies are to predict performance from chemical and physical measurements, or to use accelerating tests which attempt to simulate some of the factors involved (sometimes called 'artificial weathering'). The delivery of properties giving good performance is a function of the formulation and application methods which must be subject to quality control testing procedures. On overarching need in all testing and measurement is to understand the repeatability and reproducibility of the test method and the interpretation of the test in terms of accuracy and tolerances. This is an area sometimes referred to as 'Test Capability'. It is axiomatic that to achieve good repeatability and reproducibility, test methods must be standardised (Table 1).

A general overview of the most common test methods relevant to wood coating is given in this chapter with reference to some existing standards wherever possible. The definition of test methods is the task

Table 1 Summary of testing objectives

Coatings development (R&D)
Setting up of a coating system
Production control
Supplier qualification
Validation of marketing claims
Certification (system or product)
Comparison and selection of coatings
Defect investigation
Verification of the conformity to specific requirements

---

of various national and international standardisation committees whose activity is in constant evolution with the active collaboration of producers, consumers and research institutes. The chapter is structured in accordance with the following schema.

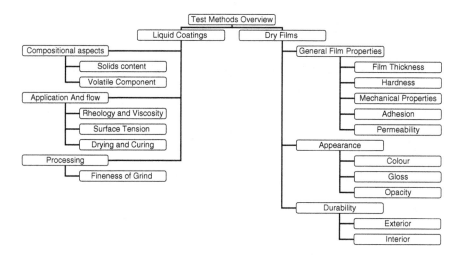

2. STANDARDS

The name standard derives from the old French word *estandar* meaning a flag, serving in the past as a reference and rallying point for military forces. The term 'standard-bearer' is still used in this context. As used here a standard is an agreed, repeatable way of doing something; this could be a test method but in many cases standards go beyond this and include Procedures, or 'Codes of Practice'. Standards can represent the internal rules of specific industrial sectors being used to simplify and clarify terminology, dimensions, technical references and performance. Official standards are prepared with the contribution of representative of the participants involved within an industrial sector, being approved by recognised organisations, working within national and international frameworks. Standards are often designed for voluntary use and do not necessarily impose any regulations. However, laws and regulations may refer to certain standards and make compliance with them compulsory. Specifications will often invoke standards as a means of ensuring conformity.

A broad objective of standards is to facilitate trade, exchange and technology transfer through:

- Enhanced product quality and reliability at a reasonable price
- Improved health, safety and environmental protection, and reduction of waste

- Greater compatibility and inter-operability of goods and services
- Simplification for improved usability
- Reduction in the number of models, and thus reduction in costs
- Increased distribution efficiency, and ease of maintenance

2.1. Standards organisations

Throughout industry test methods proliferate; many are used purely internally or for specific contractual purposes. However, in the light of the overarching objectives listed above, it is recommended that standards produced under the auspices of recognised organisations should be used wherever possible. Standards are managed by various National and International Standards Organisations, as described below, but in addition there may be scientific organisations whose activities are recognised and used globally. An example of the latter is the CIE (International Commission of Illumination) which has 38 National Committees.

All developed countries have a national standards organisation; within Europe these will include:

- BSI – British Standards (http://www.bsi-global.com/en/)
- UNI – Italian Standards (http://www.uni.com/it/)
- DIN – German Standards (http://www.din.de)
- NF – French Standards (http://www.afnor.org/portail.asp)

National Standards Organisations also cooperate with the European Standardisation Committee (CEN) which comprises the National Standards Bodies of the 30 European countries including the National Standards Bodies of European Free Trade Area countries. The aim of European standardisation is to create a uniform body of standards meeting modern needs and applying throughout the single European market. CEN also has close liaisons with European trade and professional organisations. Within CEN members, there is an agreement that European Standards will displace and supersede equivalent National Standards; publication will then be under a numbering system which includes the national and EN prefix (e.g. BS EN 927-3:2000).

Outside Europe, there are many other standards organisations. The US standardisation system includes the American National Standards Institute (ANSI) and the National Institute of Standards and Technology (NIST). ASTM International, originally known as the American Society for Testing and Materials (ASTM), was formed over a century ago, is one of the largest voluntary standards development organisations in the world source for technical standards for materials, products, systems and services (http://www.astm.org). ASTM often parallels European activity and ASTM standards are widely used outside of the USA.

Finally, there is an international organisation known as ISO. ISO is a network of the national standards institutes of 157 countries, on the basis of one member per country, with a Central Secretariat in Geneva, Switzerland, that coordinates the system. ISO is a non-governmental organisation: its members are not delegations of national governments. Nevertheless, ISO occupies a special position between the public and private sectors. This is because, on the one hand, many of its member institutes are part of the governmental structure of their countries, or are mandated by their government. On the other hand, other members have their roots in the private sector, having been set up by national partnerships of industry associations.

The relationships between the different organisations can be complex and many standards will be issued under the auspices of a national, European and international category: for example, BS EN ISO 2808:2007 (Paints and varnishes – Determination of paint film thickness). This standard is itself an example of the difficulties of standardisation as it covers 22 methods of determination and has been criticised for its complexity. However, even a simple concept like film thickness can become ambiguous when the surface is rough, or as is the case with wood some penetration takes place.

Due to the long development time of some standards, they may be circulated at the development stage in which the numbering system may have an additional prefix. European Standards may be denoted as prENs or DD (provisional, or 'draft for development').

3. CHARACTERISATION OF LIQUID COATINGS

Note. The following sections give an overview of specific test methods and at least one National or International Standard. In all cases, much detail is omitted and the appropriate source should be consulted to carry out the test.

3.1. Compositional aspects

The composition of coatings has been previously classified into four generic ingredients: 'binders', 'solvents' (volatile organic substances or water), 'additives' and 'pigments' (including colorants and extenders). To obtain general information about the nature of a coating material, two simple tests can be carried out; determination of the total solid content and determination of the pigment fraction. This gives information about the pigment binder ration (also expressed as the pigment volume concentration). Such tests also indicate the volatile component and hence the likely implications under the Solvents Emission Directive. Simple tests cannot distinguish the amount of additives present but inasmuch as these are usually present in small quantities, these two tests can give a rough indication about the general composition of a coating.

Other chemical analysis can improve with further details the nature or the specific properties of the various components.

3.2. Solid content

The solid content of a liquid coating represents the fraction in weight remaining onto the substrate, after the entire medium is evaporated. The solid content is usually reported as percentage to the total weight. The determination of such parameter is carried out by the simple heating a certain amount of the liquid coating for a defined period (typically 1 h at 105 °C). The non-volatile content of a sample is not an absolute quantity but depends on the temperature and period of heating used for the test as well as the surface area to volume ratio of that portion of the sample under test. The solid content is expressed by the following calculation:

Solid content (%) = 100 × final weight/initial weight.

The content of volatile compounds is the difference to 100 of this percentage.

Among the existing test methods is *EN ISO 3251 Paints and varnishes – Determination of non-volatile substances*.

Volume solids is the percentage residue by volume obtained when a paint film is applied at a specified thickness and dried under specified conditions. Most paints would be applied at the recommended film thickness and dried at 23 °C for 7 days.

EN ISO 3233 is the method which is used to determine the volume of dry coating obtained from a given volume of liquid coating. This is considered to be the most meaningful measure of coverage (the area of surface covered at a specified dry film thickness by unit volume of coating).

Although in principle it is possible to calculate volume solids using the masses and densities of the raw materials in a formulation, in practice this may be unreliable. This is because there is the possibility of air entrainment and chemical reaction, also the densities of the components may not be known with sufficient accuracy.

3.3. Pigment content

As pigments and extenders are generally inorganic substances, the determination of such parameter for paints can be based on the complete incineration of a portion of the liquid coating. The ash content is determined at high temperature, generally around 500 °C, weighing the sample before and after such thermal treatment. At this temperature, all the organic compounds are oxidised to CO_2.

The ash content is expressed as percentage by the following formula:

Ash content (%) = 100 × final weight/initial weight.

If organic pigments are present, the quantitative determination can be carried out by appropriate physical separation methods based on the insolubility of these substances.

An appropriate test procedure is described in international standard *ISO 14680 Paints and varnishes – Determination of pigment content*. This document contains three different methods: (1) centrifuge method, (2) ashing method and (3) filtration method. The first and the third are suitable for heat sensitive pigments (organic).

The centrifuge method utilises the principle of sedimentation depending on the different densities between the liquid and the solid fractions. The filtration method separates the particles considering their dimensions by the use of appropriate sieves.

3.4. Density

The determination of the density (mass per unit volume in kg m^{-3}) of paints is an important quality control test; density is a significant variable in respect of storage, processing and film thickness. It is through the volume term a temperature-dependent variable, and the latter must always be specified. The most common method of measuring this property is *EN ISO 2811-1 – The determination of the density of paints, varnishes and related products using a pyknometer* (otherwise known as a density cup). This method can be used for most samples provided they do not contain entrapped air in which case a pressure cup must be used (EN ISO 2811-4). Other methods that are occasionally used are *EN ISO 2811-2 Determination density by an immersed body* – a method based on the Archimedes principle and *EN ISO 2811-3 – Oscillation method*, based on the principle that the resonance frequency of a U-tube filled with paint material will vary with the density of the material.

3.5. Determination of the volatile organic compounds (or content)

As noted elsewhere in this book, legislation such as the Solvents Emission Directive places considerable emphasis on the volatile organic content (VOC) of coatings and is prescriptive on the amount of solvent that can be released into the atmosphere. Determination of the VOC is thus essential to all coating operations.

In Table 2, the solid content and the simple %VOC content for some clear coatings are listed. These values are considered at the application stage, derived from different publications. They should be considered as indicative only, as the values of real applications could be rather different depending on the wide range of formulations existing and also by the variation in the amount of thinner added by users. Paints present higher solid contents than varnishes if values are expressed as percentage: weight to weight. This is due to the presence of pigments and extenders increasing the density of paints. However, because pigments have a lower volume than solvents and binder, the net effect can be to increase the volatile content depending on the definition of VOC that is used.

Table 2 VOC of some coatings

Coating material	Solid content (%)	VOC (%)
Solvent-based stains	1–10	99–90
Water-borne stains	1–20	3–10
Solvent-based impregnating stains	10–20	90–80
Water-based impregnating stains	10–20	2–8
Alkyd coatings	30–50	70–50
Cellulose nitrate coatings	15–25	85–75
Ureic coatings	25–40	75–60
2K polyurethane coatings	30–40	70–60
2K HS polyurethane coatings	45–60	55–40
2K acrylic coatings	15–30	85–70
PEPO polyesters	60–85	40–15
UV polyesters	60–95	40–5
UV acrylic coatings	60–99	40–1
Water-borne coatings	30–40	3–10
Powder coatings	99–100	0–1

VOC is defined in several different ways not all of which are covered by legislation. Some are defined by pressure groups or commercial organisations to their suppliers. There is also a fundamental difference between the USA and Europe in that the former concentrates on photo-chemistry and the latter focuses on volatility as the basic criteria for definition, for example:

- A chemical which is photo-chemically active and contains carbon
- Any organic compound with, at normal conditions for pressure, a boiling point lower than or equal to 250 °C

This means that in the USA (depending on the State), some solvents such as acetone are 'exempt' and need not be included in the VOC calculation. Whatever the definition it must then be expressed in appropriate units, for example, gram per litre, even this requires a definition of what is to be included, some regulations do not count the water content, thus effectively increasing the VOC of water-borne paints but reducing the use of water simply as adulteration.

In principle, the VOC of a coating can be calculated from knowledge of the formulation, and in the case of coatings for which the volatile compounds are represented almost exclusively by organic solvents, the determination of the solid content allows an estimate of the solids content to be made. However, this may not be allowed by legislation and for some coating types there are other complications as noted below.

In general for conventional solvent-borne paints, test method ISO 11890-1 is used. The non-volatile matter of the paint is determined in accordance with ISO 3251 and density in accordance with ISO 2811-1. The water content of the paint may then be determined using a titration technique employing a Karl Fischer reagent in accordance with ISO 760. A calculation is then performed to give the VOC content in $g\ l^{-1}$ of the sample calculated according to the prevailing regulations.

For water-borne paints, the volatile organic components are separated by a gas chromatographic technique in accordance with ISO 11890-2. Depending on the sample type either a hot or cold sample injection system is used, but hot injection is the preferred method. After the compounds have been identified, they are quantified from the peak areas using an internal standard. A calculation is then performed to give the VOC content of the sample. Gas chromatography can be useful also in the case of solvent-based coatings as, identifying each single component, allows enabling the possible exclusion of solvents not classified as 'volatile organic compounds' depending on the definition considered.

There are several units used to express VOC according to the industry. In the decorative market, the units are gram per litre ($g\ l^{-1}$) and relate to the product itself. Elsewhere units may relate to the environment and be expressed as a mass concentration $mg\ m^{-3}$, or volume concentration as $ml\ m^{-3}$ or ppm. For industrial processes in conditions of continuous emission, a mass flow ($g\ h^{-1}$) may be measured. The latter measurements provide the necessary input for a solvent management plan (SMP) which must be used to account for the difference between purchase and consumption of solvents during a process. There are also a number of possibilities for process engineering solutions to emission reduction such as afterburning.

It should be noted that for specific sectors, such as the furniture industry, the implementation of legislation may follow a different timetable within European Countries and some differences according to the size of the operation. Most will now require a SMP to be in place with targets for reduction in the future.

3.5.1. Photo-curing coatings
In the case of photo-curing coatings, reactive diluents partially react with the resin during the drying process becoming part of the coating film. The remainder evaporates together with any other solvent.

The true solid content of such coatings is thus represented by the sum of the weights: of pigments, if present, of the resin and of the fraction of reactive solvent finally bonded to the resin. Such value strongly depends on the process utilised, the most critical phases being represented by the application and the flash off periods. Depending on the monomer type, quantity and also on the temperature and duration of such processes, evaporation can take place at different levels.

The standard methods described in Section 3.4 are not suitable for photo-curing coatings, as the monomers could give false results.

American standard ASTM D5403 *Volatile content of radiation curable materials* describes a test method usable for the determination of the solid content of such products but excluding, from the application field, possible volatile monomers as styrene. The presence of volatile monomers strongly affects the repeatability and reproducibility of the test. A standardised method applicable to all the photo-chemical curing products does not presently exist.

Only the execution of direct tests in real conditions could give sufficiently reliable results for the specific system considered.

3.5.2. Chemically curing polyesters

Unsaturated resins prevalently dissolved in styrene constitute chemically curing polyesters. The curing process is promoted by the addition of a metal salt and peroxide. As with photo-curing, during the drying phase, a portion of the reactive solvent reacts with the resin while the rest evaporates together with the other volatile organic compounds. The determination of the solid content by simple heating the liquid coating would not be appropriate, as the chemical reaction of styrene would not be considered. The real solid content depends by the specific conditions adopted (application rate, drying conditions and time). Again a standardised method does not exist and only 'in situ' tests could be able to give some realistic results.

3.6. Other test methods

There are many other test methods used by the coatings industry which are outside the scope of this book. Some of the most common tests are listed below and may be cited elsewhere; further information may be obtained from the bibliographic sources at the end of the chapter:

- Chemical analysis:
  - Acid value (EN ISO 3682)
  - Hydroxyl value (EN ISO 3682 and 4629)
  - Number of NCO groups (ISO 10283, DIN 53241-1)
  - Amine or acid equivalents (MEQ_B, MEQ_S) (EN ISO 3682)
- Instrumental analysis:
  - Spectrographic
  - Infrared spectroscopy for the identification of the binder (ASTM D2621)
  - Chromatographic
- Flash point representing the lowest temperature at which the solvent vapours are capable to be ignited (EN 14370 closed cup method)
- Surface tension (EN 14370)
- Water content of solvents and thinners (ASTM D1364)

4. PROPERTIES RELATED TO APPLICATION

4.1. Viscosity

Viscosity is the most important property of liquid coatings related to application, flow and drying processes. It is effectively the resistance of a liquid to deformation increasingly with increasing rate of deformation, in the simplest terms it is the shear stress divided by the shear rate. The scientific study of deformation and flow is known as 'rheology'.

Inappropriate viscosities or rheological behaviour can cause several problems. Good levelling can be resisted by high viscosity and also the necessary escaping of entrained air bubbles can be prevented leading to the formation of 'pinholes'. Other defects in the film can arise due to an imbalance between surface tension gradients and viscosity. These include a mottle known as 'orange peel' and various colour separation effects.

Viscosity is an important application parameter and the requirement for a higher viscosity will generally increase in the general order:

flow coating < pneumatic spray
< air-less spray < curtain coating < roller coating.

Viscosity is also important for pigment dispersion and must be adjusted according to the manufacturing method used.

By definition, the viscosity of Newtonian fluids is a material constant depending for its value on temperature, pressure and concentration (thermodynamic variables). Thus if the viscosity is plotted as a function of shear rate, it will show a horizontal straight line. Very few of the materials and compositions used in paint would show such behaviour because the viscosity is not a constant, but a function of the shear rate and known as non-Newtonian. When viscosity falls as a function of increasing shear rate, the behaviour is best described as 'shear thinning', though the term 'pseudo-plastic' is sometimes used. Time-dependent shear thinning, that is, to say a fall in viscosity at a given constant shear rate, is properly described as 'thixotropy'. This is often confused with shear thinning. Materials that are thixotropic will also show shear thinning if the shear rate is raised, and then show further thixotropic viscosity loss if held for a period of time at the higher shear rate. Recovery of viscosity after shearing is useful behaviour as it enables good initial flow followed by a period where sagging (curtaining) will be resisted.

Due to this complex behaviour, the viscosity of coatings is not easily measured and the shear rates, temperature and shear history need to be specified. There are many different types of viscometer which range from simple devices with a fixed shear rate, to complex instruments where the shear rate or shear stress can be controlled or varied. In general, the more simple types are used for quality control and the more complex for R&D purposes.

Types of viscometer include:

- Capillary viscometers
- Cone and plate
- Rotating disc
- ICI rotothinner
- Mixing rheometers
- Torque methods
- Bubble viscometers
- Efflux cups
- Paddle viscometers
- Extensional viscometers
- Oscillatory methods

One of the simplest examples of a low shear rate instrument is the efflux flow cup which is a defined volume and shape with a hole in the bottom through which the paint flows. Such cups are typically used to check paint consistency before spraying. Use consists of filling the cup with paint and measuring the time taken for the paint to flow through. A large number of national flow cups exist – standards include ASTM D1200, DIN 53 211 and BS 3900:A6. Efforts have been made to encourage the uptake of the international – EN ISO 2431 series of cups but while its use is increasing the national varieties still persist in quite large numbers.

Instruments for the measurement of viscosity at low-intermediate levels of shear include most rotational viscometers where a paddle or spindle is dipped into a container of the paint material. There is usually a direct readout of the paint viscosity in Pascal seconds. Examples include the Brookfield instrument (ASTM D2196), the Rotothinner (ISO 2884-2) and the Krebs Stormer viscometer (ASTM D562). The rotational speed of the Brookfield may be varied so that the shear thinning behaviour of the paint, including thixotropy, can be studied.

An example of a viscometer with high shear is the cone and plate viscometer (ISO 2884-1). The method uses a cone and plate viscometer which consists of an electrically driven shallow cone the vertex of which touches a rigid temperature-controlled plate. The test liquid fills the narrow gap between the cone and the plate and the torque is measured either electronically or mechanically. The choice of cone geometry and speed or rotation enables a shear rate of between 9000 and 12,000 s^{-1} to be attained and this high shear rate corresponds to that produced during brushing, roller coating and spraying operations.

4.2. Pot life

In the case of coatings that dry only by physical evaporation, any increase in viscosity during the application process is only due to the evaporation of the volatile solvents. Restoring the viscosity to the required value can usually be carried out by the simple addition of thinner, and in closed containers the material should have a long useable life. For chemically curing coatings, viscosity change arises also by an increase in the molecular weight of the binder as a consequence of the chemical reactions involved. Restoring of the required viscosity by simple addition of thinner is not a feasible operation as the chemical and physical properties of the liquid coating are definitely and irreversibly altered. The 'pot life', known also as 'working life' or 'usable life', represents the length of time a two-pack coating retains a viscosity low enough to be suitable for the application, after the components are mixed. The conventional pot life is defined as the time in which the viscosity doubles its initial value. Pot life depends on the ambient temperature. The most common reference standard is the *ISO 9514 Paints and varnishes – Determination of the pot life of multi-component coating systems* (Table 3).

4.3. Minimum film formation temperature

One of the most important properties of coating dispersions (e.g. an acrylic latex) is the minimum temperature at which a continuous film can be produced during the drying phase. This process, called 'coalescence', is described in Chapter 10. The following standard describes a test method capable to determine the temperature below that, cracking or whitening effects can be detected in the un-pigmented film. A reference method can be found in the standard *ISO 2115 Polymer dispersions – Determination of white point temperature and minimum film-forming temperature.*

4.4. Drying time

'Drying time' is the period between the application of the liquid coating and the formation of a solid film capable of resisting some deformation (such as being touched as in 'touch-dry') or resistance to solvent. The end point for a test is often a combination of curing (cross-linking) as well as drying. Because of the different methods of film formation between

Table 3 Relative comparison of the pot life of three different families of chemically drying coatings

Coating family	
Chemical curing polyesters	Short pot life
2K polyurethane coatings	⇕
Ureic coatings	Long pot life

technologies (Chapter 10), the end point of drying test is seldom unequivocal, and it is important to specify conditions. A distinction is often made between 'surface' and 'through' drying.

ISO 1517 is used for the determination of the *surface* drying characteristics of a paint or varnish film which dries by reaction with air or by a chemical reaction of its components. This method gives a measure of the time that elapses between the application of a coating and that moment when the surface first becomes sufficiently dry that material such as Ballotini beads, no longer adheres to it. ISO 9117, in contrast, is the method used for the determination of the *through* drying characteristics of a paint; it deploys a machine which presses a gauze fabric covered plunger against the film and then rotates it through 90°. This method is sometimes called a 'mechanical thumb'.

There are also devices which follow the progress of drying by using a mechanical stylus. ASTM D5895 describes mechanical devices for recording the drying time of organic coatings. The method employ a stylus moved along a drying paint film at a constant speed. When the coating is dry, it is usually possible to determine the various drying stages by the characteristic pattern left in the coating by the stylus. ISO 3678 assesses the resistance of a coating film to imprinting by a nylon gauze under a specified force applied for a specified time. The method may be carried out either as test by determining whether the print-free state has been reached after a specified period of drying or, in the case of stoving coatings, after stoving and ageing under specified conditions, or by repeating the print-free test at suitable intervals.

5. APPLIED COATINGS (DRY FILM)

5.1. Properties related to appearance

Most wood coatings, in addition to any protective function, are expected to enrich the surfaces to which they applied with characteristics which will include:

- Transparency
- Colour
- Gloss (or sheen)

These properties arise from the interaction of light with matter but combined with a subjective interpretation in the brain. Although colour may be used in some decorating applications in a monotone hue and with diffuse spectral reflectance, it is often the highlights or directional reflections which bring the surface alive. Sensations such as 'gloss' will share with colour a physical, physiological and psychological element. Human observers may judge glossiness by the distinctness of reflected images whereas instruments (glossmeters) are usually measuring the intensity. Furthermore, light reflection will depend on the size surface irregularities as a function of the incidence angle and must be specified accordingly. Interactions between colour and shape are important aspects of recognition and differentiation. For example, the perceived difference between a plastic and metal object is based on the sharpness and colour of the highlights.

In most paints, it is desirable to obliterate the substrate and hence good opacity, or 'hiding power' is required. Wood is aesthetically attractive in its own right and this leads to coatings with various degree of transparency (varnishes, lacquers, stains, lasures) in which the final appearance combines attributes of the coating and the substrate. In the following paragraphs, the interaction of light with coated surfaces is considered starting from the general concept that when light hits an object it can be transmitted, reflected, refracted or absorbed. In a very simplified manner, it is possible to assume that transmission is associated with transparency, reflection with gloss and absorption with colour sensations.

5.1.1. Light transmission (transparency and hiding power)

Transparency is a property related with varnishes and conversely hiding power is a property of paints.

Apart from the definition given by EN 927-1, there are not specific methods used in the wood sector to measure the coating transparency. Very often, this evaluation is part of the overall visual sensation deriving by a direct view of the coated samples.

TRANSPARENCY DEFECT: BLUSHING

A possible defect in coating transparency is termed 'blushing'.

It is defined as a milky opalescence that sometimes develops especially during drying. It may derive from different factors:

- Condensation of moisture
- Precipitation of some of the solid components
- Inclusion of solvents or water in the dry film
- Partial detachments of the coating layers

Hiding power is defined as the ability of paint to obscure the surface over which it has been applied. It is usually reported as the minimum film thickness obscuring completely the substrate or by the surface area that can be completely obliterated by a defined amount of coating. There are different test methods for measuring the hiding power of paints.

The simplest methods are based on the application of the coating product onto particular black and white substrates (contrast test charts). The result is expressed by the minimum film thinness necessary to hide the black and white pattern to the view. This measure can be carried out also by spectrophotometers reporting the reflectance of the two areas (black and white). Hiding power strongly depends on pigment content, dimension (particle size distribution) and type. The higher the scattering power and the light absorption, the higher is the contribution of the pigment to the hiding power.

White pigments achieve opacity primarily through scattering, whereas with coloured pigments the contribution is from absorption. In the case of normal TiO_2 maximum scattering and whiteness is the objective. There will be an optimum particle size (around 0.2–0.3 µm), for maximum scattering, but this maximum will also depend on refractive index and wavelength. Smaller particles, in the nano-size range, will be transparent. With absorbing pigments (i.e. coloured) a decrease in particle size will increase colour strength (the ability to tint) but decrease opacity.

The Kubelka–Munk equation is used in many standardised methods, and represents a phenomenological basis for the treatment of absorption and scattering characteristics of pigments.

The basic K–M equation is

$$K/S = (1 - R_\infty)^2/2R_\infty.$$

where R_∞ is the reflectivity, K is the absorption and S is the scattering coefficient.

There are several variants of the methods used to determine hiding power in conjunction with the Kubelka–Munk constants (see Paint and Coating Testing Manual, 14th Ed., ASTM 1995 for a useful review).

The most common standards related with the hiding power are the following:

- *ISO 6504-1 Paints and varnishes – Determination of hiding power – Part 1: Munk method for white and light-coloured paints*
- *ISO 6504-3 Paints and varnishes – Determination of hiding power – Part 3: Determination of contrast ratio of light-coloured paints at a fixed spreading rate*
- *ISO 2814:1973 Paints and varnishes – Comparison of contrast ratio (hiding power) of paints of the same type and colour*

5.1.2. Light reflectance (gloss)

The fraction of light reflected by surfaces is called reflectance. The amount of reflected light depends on the nature of the substrate and the incident angle of the rays. The way in which the rays are reflected is associated to the perception of shininess and brilliance.

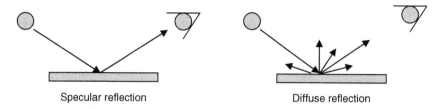

Specular reflection   Diffuse reflection

Figure 1   Representation of light reflection.

The reflected light may be concentrated at angle numerically equal to the angle of the incident ray, or scattered in all directions. Between these two extremes called: 'specular' and 'diffuse' there are many possible intermediate distributions. The directional reflecting properties of a surface are comprehended in the term 'gloss' which definition is given by the international standard ISO 2813: 'Visual impression deriving from the reflection properties of the incident light onto a surface of a coating material' (Fig. 1).

The factors affecting gloss are the refractive index of the material, the incident angle of light and the surface topography. Materials with smooth surfaces appear glossy, while rough surfaces, in micro-metre range, not reflecting light specularly, appear matt. Specular reflection is measured with glossmeters. White light is concentrated by a lens system onto the substrate. The reflected beam is then measured through a photo-detector. The common angles of incidence for gloss measurement are 20°, 60° and 85°. Low gloss surfaces are recommended to be measured with 85° settings. *ISO 2813* describes a method for measuring specular gloss of non-metallic paint films and *EN 13721* specifies a test method for furniture surfaces. Surface roughness at different size levels is responsible for alterations in the distinctiveness of the reflected imagine especially in glossy coatings. The most common terms related with such perceived reflectance defects are haze and levelling defects including 'orange peel':

- *Haze*. Haze is defined as the blurring of the outlines of reflected images by the formation of scattering particles on or just beneath the surface of the film.

  The surface appears milky or cloudy. Haze is measured by the intensity of the transmitted light that is scattered more than 2.5°. There are specific instruments determining this parameter based on standardised methods like *ISO 1380 Paint and varnishes – Determination of reflective haze on paints films at 20°*.

- *Levelling defects*. Most coatings are not flat after application but will contain irregularities such as bush marks, roller striations, spay mottle. Provided viscosity is not too high then surface tension will bring about levelling. Gravity plays no part in levelling but can cause sagging at higher film weights. According the Orchard equation [1], levelling is most rapid when the wavelength of the defect is small and film thickness is high. As the viscosity increases during drying, the point is reached where defects become trapped.

- *Orange peel*. This is a defect, which imparts to the film the dimpled appearance of the surface of an orange in consequence of the 'wavy' structure of the surface; it is associated with spray application and is in part due to the more mottled surface achieved by spray methods. However, orange peel can appear on the surface of a glossy coating that was initially smooth on application. It is particularly associated with solvents that have fast evaporation rates and with electrostatic spraying. It arises from a surface tension differential (which might be caused by evaporation or electrostatics) and can be reduced by lowering the dynamic surface tension, for example, with silicone fluids and other specific surfactants. However, care is needed to avoid introducing other surface tension driven defects.

Orange peel appearance can also be associated with other factors (substrate porosity, substrate humidity, appropriate cooling of the coated panels at the end of the process, etc.).

5.1.3. Light absorption (colour)

Pure 'white' light is colourless and colour might therefore be described as the non-white distribution of light. This will arise from the interaction of light with matter in the presence of an observer. Excitation of electrons will cause selective emission or absorption of light which may also be reflected or scattered. The term 'hue' is defined as an attribute of a visual sensation according to which an area appears to be similar to one, or to proportions, of the perceived colours red, yellow, green and blue.

Colour is an important property of every surface contributing to the overall appearance of a finished product. Coatings play an important role conferring the colour to a substrate; therefore, this subject represents one of the most important topics for both manufacturers and users of coatings.

Colour is the human perception of the wavelength of light depending upon three fundamental elements:

(1) The light source
(2) The object
(3) The observer

The first element is represented by a source of electromagnetic waves in the range between 400 (blue/violet) and 780 (red) nm. This range is called the visible range. The most common light source is sunlight. Other light sources (bulbs, fluorescent tubes, etc.) differ in their emission spectrum (a plot that records the intensity at each wavelength). The waves emitted by the light source hit the object being selectively absorbed by its surface. The reflected rays reach the observer and the combination of wavelengths (reflected spectrum) is transformed by the human brain into a colour sensation. If an object does not absorb any of the visible radiations, it appears 'white' and conversely, the complete absorption is perceived as black. All the other colours are consequence of partial absorptions in this spectral range.

The physiology of the colours perception by the human eye is based on a complex combination of retinal receptors (rods and cones) and some nerve fibres sensitive in three different wavelength ranges. The incoming spectrum is thus effectively reduced by the eye into three stimulus or values representing the intensity of the response of each cone type. The combination of the three stimuli produces to the brain the perception of colours in terms of brightness, hue and colourfulness [2].

'White' light contains all wavelengths and can in principle be synthesised by projecting a suitable blend of coloured 'primary' wavelengths (i.e. from the violet 380 nm to the red 780 nm region of the spectrum). Conversely if wavelengths are removed from white light, the remainder will be coloured. Thus, a dye which absorbs red light will appear blue green in transmission. The two extremes are known as additive and subtractive colour mixing (Table 4).

In reality pure 'primary' colours do not exist and the coatings industry must use a wider gamut of coloured pigments.

5.1.3.1. Colour measurement   For an objective evaluation of the colours, it is initially important to use a standardised light source. The most common is represented by the so-called D65 illuminant. It is defined as the emission of a black surface heated at the temperature of 6,500 K which spectrum is comparable to that of daylight (D = daylight). The light reflected by the object is then evaluated by special detectors simulating the human perception.

Some experiments directly carried out on population and elaborated by the Commission Internationale de l'Eclairage have provided the basis for the definition of a numerical system allowing the expression of colour by means of three values (X, Y and Z) related to the human perception. The CIE chromaticity diagram is usually shown as a two-dimensional representation of chromaticity coordinates. This form of colour space has certain limitations; it is based on relationships between equal colour

Table 4 Additive and subtractive colour mixing

Additive mixing	Subtractive mixing
Red + green = yellow	Yellow + magenta = red
Green + blue = cyan	Yellow + cyan = green
Blue + red = magenta	Cyan + magenta = blue
Red + green + blue = white	Magenta + cyan + yellow = black

stimuli and not colour perception. It can be used to show if two colours have the same chromaticities but not what they would look like to an observer, and in particular it cannot show in a systematic manner how two non-matching colours differ. Unlike other systems, there is no unique position for black in the CIE chromaticity diagram. Notwithstanding these limitations, the CIE diagrams offer many insights into colour, including the gamut that can be achieved with specific sources, and is especially useful in colour television.

In the course of time, this system has evolved leading to the well-established colour system called CIELAB.

This practical methodology allows defining every colour as a single point in a Cartesian trichromatic space where the coordinates are represented by the axis:

- L^* (white–black)
- a^* (red–green)
- b^* (yellow–blue)

These coordinates derive by mathematical elaborations of the values X, Y and Z having the advantage to allow simple and effective colour determinations.

One of the most useful aspects of such system is the opportunity of representing the difference between two colours as a difference in the values of the three coordinates (ΔL^*, Δa^* and Δb^*).

Another option is to express such difference as the geometric distance ΔE^* being the length of the segment between the two points in the trichromatic space:

$$\Delta E^* = \left[(\Delta L^*)^2 + (\Delta a^*)^2 + (\Delta b^*)^2 \right]^{1/2}.$$

This system is useful in defining colour tolerances and in practice is complimented by various colour order systems. Colour order systems (colour collections, colour cards, etc.) serve a practical need in providing a common vocabulary between customer and supplier. They have a long history and many will predate the CIE colour spaces. Colour order systems will often be based around specific materials or the requirements of particular organisations and will have inherent differences. Colour order systems can be empirically, or colorimetry based. Examples include:

- The Munsell system groups colours by hue, lightness and chroma (saturation), the perceptual differences between two neighbouring samples is made as constant as possible.
- The New Colour System (NCS) system used in Sweden and elsewhere has the three variables in the colour atlas of blackness, chroma and hue. There is no simple relationship between NCS and CIELAB systems.

- The DIN colour system [3] is linked to the CIE system and uses black, hue and saturation as variables, with extended use of colorimetry for interpolation and extrapolation. It finds most use in Germany.
- The Optical Society of America OSA system; where primacy is given to producing a system in which the distance between two points represents as closely as possible the perceptual size of the difference between the equivalent samples. There is no correlation with hue, chroma or saturation; instead, three numbers represent lightness, yellowness and greenness.
- The RAL-DS design system which employs CIELAB as an order system.

METAMERISM

Metamerism is a phenomenon where coloured materials seemingly the same viewed under a given light source appear different under different light sources. For example, two colour samples might appear the same in natural light, but not in artificial light. This means that even where the samples seem to match they have a different spectral composition. The greater the difference in spectral composition, the more they are likely to seem different if the lighting or observers are changed.

The CIE recommends calculation of an 'Illuminant Metamerism Index' consisting of the size of the colour difference between a metameric pair caused by substituting a test illuminant for a reference illuminant of different spectral composition.

Two standards related with the metameric evaluation of paints are:

(1) *ASTM D4086a Standard Practice for Visual Evaluation of Metamerism*
(2) *ISO 3668 Paints and varnishes – Visual comparison of the colour of paints*

5.1.3.2. Colour measuring instruments   All colour measurement must adopt a standard illuminant, a standard white (for calibration) and a standard geometry for illumination and viewing conditions. In many cases, these will be integral to the chosen instrument.

Two different types of instrument are used for measuring colour (or rather the physical aspects that give rise to the phenomenon of colour); they are:

(1) Colorimeters
(2) Spectrophotometers

Colorimeters use tristimulus broad band absorption filters. The colorimetric data correlate with human eye–brain perception. Colorimeters have set illuminant and observer combinations. They are primarily for routine inspection, quality control and the adjustment of small colour differences.

Spectrophotometers provide wavelength-by-wavelength spectral analysis using narrow band gratings or interference filters. They are more complex that colorimeters and can be used for colour formulation, measurement of metamerism, and R&D as well as quality inspection. They offer a range of illuminant observer combinations covering directional and diffuse conditions. Forty-five degree illumination is common but there are many CIE geometries suited to specific purposes.

45/0° Geometry. The light source is placed at 45° and the detector (observer) perpendicular to the plane (0°). A variant of such system is to invert the two angles (geometry 0/45) (Fig. 2).

Diffuse geometry. These spectrophotometers irradiate the object with a diffuse light in consequence of the presence of a suitable sphere which internal surface is perfectly white. The detector angle is usually placed at 8° to the plane (geometry d/8). These instruments are able to measure also light reflectance. Spectrophotometers with diffuse geometry are also able to evaluate metameric effects (Fig. 3).

Most surfaces reflect some light that has not penetrated the surface and is therefore not affected by the colorant, it is for this reason that matt surfaces do not appear as saturated as glossy surfaces. When measuring it is necessary to specify whether the specular component should be included (SPIN) or excluded (SPEX).

More information about colour and its measure can be found in the following standards:

- *ISO 7724 – Part 1: Paint and varnishes – Colorimetry – Principles*
- *ISO 7724 – Part 2: Paint and varnishes – Colorimetry – Colour measurement*
- *ISO 7724 – Part 3: Paint and varnishes – Colorimetry – Mathematical formulae used to calculate colour differences in terms of lightness, chroma and hue*

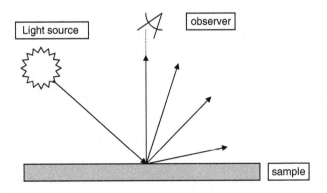

Figure 2   Representation of a 45/0° geometry spectrophotometer.

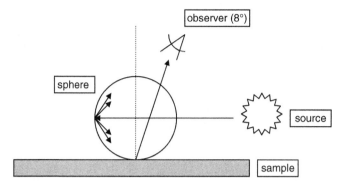

Figure 3 Representation of a diffuse geometry (*d*/8) spectrophotometer.

ISO 3668 describes a method of carrying out a visual comparison of the colour of paint films or related products with a standard colour under natural or artificial lighting conditions.

Visual colour matching has the advantage that it provides information on how the colour is perceived. The disadvantage of visual colour matching, however, is that subjective may well result in disagreement between observers.

6. COATINGS PERFORMANCE

6.1. General properties

6.1.1. Film thickness

Film thickness is one of the major determinants of coating performance and appearance. Film thickness is not an unequivocal measurement and there are many descriptors which take into account distributions (profiles) of film thickness as well as the average film thickness. *EN ISO 2808 Paints and varnishes – Film thickness determination* is an international standard that describes many of the methods that are applicable to the measurement of the thickness of coatings applied to a substrate. Methods for determining wet film thickness, dry film thickness and the film thickness of uncured powder layers are described. Figure 4 gives an overview of methods that are used.

A microscope may also be used to measure film thickness. Apparatus known as a paint inspection gauge (PIG) cuts a wedge-shaped cut into the paint film. The cut is then examined under a portable microscope and the width of the cut is measured with a graticule. As the angle of the cut is defined by the blade of the instrument, it is possible to convert the apparent width of the cut into film thickness by trigonometric functions.

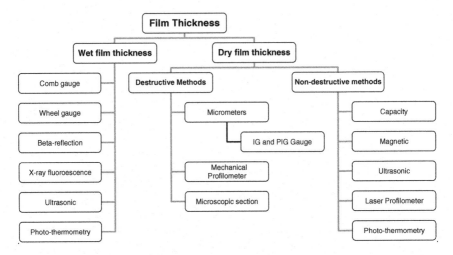

Figure 4 Methods for determining film thickness.

Although the test is destructive, it does have the advantage that if the paint film is comprised of several discernable layers the thickness of each layer may be determined.

THEORETICAL DRY FILM THICKNESS

It should be noted that many wood coatings penetrate into the substrate and the concept of film thickness becomes ill defined. It may be useful therefore to also consider the theoretical film thickness.

The theoretical wet film thickness is obtained by the simple expression:

$$\text{Wet film thickness } (\mu m) = 1000/\text{spreading rate } m^2\, l^{-1}.$$

The theoretical dry film thickness may be obtained from the fractional volume solids of the paint. Thus, a litre of paint of 50% solid content spread over 10 m^2 will have a wet film thickness of 100 μm and a dry film thickness of 100 $\mu m \times 0.5 = 50\ \mu m$.

The required spreading rate in $m^2\, l^{-1}$ to achieve a required dry film thickness is calculated by multiplying the % volume solids by 10 and dividing by the desired film thickness in μm. For example, given that the % volume solids is 35% and required dry film thickness 25 μm, then required spreading rate $= (35 \times 10)/25 = 14\ m^2\, l^{-1}$.

Note. 1000 μm = 1 mm. The term micro-metre may be preferred to micron.

The term 'build' is sometimes used synonymously with film thickness, although sometimes the term is used to describe the apparent, rather than actual, film thickness. See EN 927-1 for current definitions in relation to wood coatings.

6.1.2. Adhesive performance

Adequate adhesion is a fundamental requirement for good coating performance and will affect all resistance and durability properties. There are two quite different aspects of adhesion. One relates to the atomic or molecular forces acting across an interface, such forces are not easily measured. The other is a measure of the forces needed to separate surfaces (thus normally a destructive test), this aspect is best described as 'adhesive performance'.

Adhesive performance is dependent on the circumstance in which it is measured and, in practice, it depends on the manner and direction in which the coating is stressed. An important factor affecting adhesion is the internal tension of the film deriving from its volume contraction due to drying and curing. Loss of volatiles involves considerable shrinkage; moreover, the cross-linking mechanisms of chemical drying coatings reduce the free volume further increasing internal stress. Thick films will be more stressed than thin films.

In the case of wood dimensional variation of the substrate, in response to humidity exchange, plays an important role in causing the coating to become strained. The internal stresses so generated are often a determining factor for the origin of spontaneous detachments the coating films.

Adhesion is often reduced in wet conditions, especially for coatings used outdoors. Typical defects deriving from adhesion failures are flaking and blistering. These phenomena are more frequent with water-borne coatings as water can act as a plasticiser, or weak boundary layer, reducing the adhesion forces between wood and substrate.

One of the most common test methods to investigate adhesion performance is described by the *ISO 4624 Pull-off test for adhesion* where the stress is applied perpendicularly to the coated surface. Another method is contained in the *ISO 2409 Paint and varnishes – Cross-cut test*. However, although loss of adhesion is often a cause of detachment, wood itself may fail in cohesion, particularly if the surface is denatured. It is always useful to examine the locus of failure as well as the magnitude of removal (Fig. 5).

6.1.3. Surface hardness

Hardness can be defined as the ability of a coating film to resist indentation or penetration by a solid object (EN 971-1: 1966). Hardness is also related to the dying state of the coating depending on solvents retention and, for chemical drying coatings, on the cross-linking effectiveness. The efficiency of the drying systems is then fundamental to achieve an adequate hardness.

Figure 5   Cross-cut test (ISO 2409).

Hardness of a coated surface is a significant parameter determining the behaviour of the coated product during its transport, manufacture and use. A soft surface can be detrimental and prone to marking by packaging or general handling. Common operations carried out at the end of the coating process (e.g. stacking of components) or during transport can cause problems.

Hardness of the surface also depends to some extent on the mechanical behaviour of the substrate. Hardness influences many different properties of the final product as the resistance of the surface to impacts, scratches and abrasion effects. Some specific test methods are described in the section related to the properties of coatings for interior use.

ISO 2815 describes a method used to assess the resistance of a dry film of paint, varnish or related product to indentation by a weighted metal wheel. The apparatus used in this test is known as the Buchholz indenter. It consists of a sharp-edged wheel (made of hardened tool steel) which is mounted in a rectangular block of metal fitted with two feet.

ISO 1522 is a damping method used to assess the hardness of a dry film of paint, varnish or related product by measuring how it reduces the oscillation amplitude of a pendulum.

There are two types of apparatus defined in this standard:

(1) The Konig pendulum
(2) The Persoz pendulum

Both pendulums have spherical balls which rest on the coating under test and form the fulcrum. Both employ the same principle, that is, the softer the coating the more the pendulum oscillations are damped and the shorter the time needed for the amplitude of oscillation to be reduced by a specified amount.

ISO 15184 describes a method which is used to assess the hardness of organic coatings using a series of pencils with leads of known hardness.

Artists' pencils are made with leads that range in hardness from 9H (very hard) to 6B (very soft). The test equipment consists of a set of 17 pencils together with a holder which allows a pencil to be held at a 45° angle while being pushed firmly across the surface of the coating under test.

6.1.4. Stackability (blocking)

One of the properties related to the surface hardness is the need to stack components after a reasonably short time after the coating process. Components should not adhere sufficiently to cause damage on separation either at this stage, or during transport. The effect is exacerbated by higher temperatures caused, for example, by the effects of sunlight.

The defect caused by a poor stackability is also called blocking being defined as the unwanted adhesion between two coated surfaces when they are left in contact under load. This term is more likely to be used in service conditions, and applies, for example, to the possibility of window frame adhering in the rebate.

Blocking is more frequently tested in the case of coating systems for exterior wood due to the typical thermoplastic behaviour of the binders used. Formulation of these products must be balancing thermoplastic properties with against operational requirements.

A standard test for stackability is *ISO 4622 Paints and varnishes – Pressure test for stackability* other optional test for blocking have been devised and can be agreed between supplier and vendor.

6.2. Coatings for exterior use

The properties required for a coating system intended to be used in exterior conditions are related to the protection of wood substrates against the factors leading to its degradation while retaining good appearance. Test methods can be divided into three broad groupings:

(1) Natural weathering
(2) Artificial 'weathering'
(3) Predictive tests (based on physical or biological properties)

The most appropriate tests established are those reproducing weathering processes based on natural or artificial exposures. Their aim is to combine and verify the effects of more degrading agents.

Other test methods presented here are related to specific properties of coating systems.

6.2.1. Weathering methods

6.2.1.1. Natural weathering exposure   Test methods based on natural exposures represent the definitive standard as they correspond most closely with service life conditions. Despite this unquestionable advantage, natural weathering exposure is usually long term with questionable repeatability and reproducibility. These difficulties arise in part from variability in weather itself, which cannot be controlled, but also the absence of agreed dosage metrics for quantifying such variation. Wood also raises problems due to the significant variability within and between species.

Weathering studies form part of the overarching discipline of service life prediction (SLP) which has been categorised into three main methodologies [4]:

(1) Descriptive
(2) Scientific measurement
(3) Reliability based

ISO 2810 outlines the general principles for determining the resistance of all types of coatings to natural weathering including the measurement of climatic factors The European standard *EN 927-3 Paints and varnishes – Natural weathering test* gives an example of a natural weathering method devised specifically for wood. The exposure is carried out for 1 year with the samples exposed with an inclination of 45° towards South (Fig. 6).

The use of a standard coating system called Internal Comparison Product, with the function of indicator of the climatic conditions during the exposure period, is an attempt to quantify the problem of repeatability and reproducibility mentioned above.

Figure 6   Natural weathering according to EN 927-3 (picture from CATAS spa).

Figure 7 Two different coatings after 1-year exposure according to EN 927-3 (picture from CATAS spa).

The mandatory parameters used for the evaluation of the surface degradation are (Fig. 7):

- Blistering
- Cracking
- Flaking
- Chalking
- Loss of adhesion
- Colour change
- Loss of gloss
- Mould growth

Other optional tests methods such as blocking and stackability are also described. Results are interpreted (see *EN 927-2*) by a scoring system which is related to end-use categories.

6.2.1.2. Artificial weathering Artificial weathering methods comprise a source of radiation energy combined with water spray and a system able to produce water condensation on the surface of specimens. A general sub-division of such systems is based on the lamp type used. There are two general families using different construction principles:

- The first is that using filtered xenon lamps. They substantially reproduce the global sunlight spectra including UV, visible and IR radiations. The equipment also allows for a greater degree of control of temperature, humidity and radiation dose than is possible with other forms of accelerated weathering. The main disadvantages of the xenon arc method are:
  - It is one of the most expensive weathering tests.
  - The test samples have to be flat and there are limitations on the size of samples that can be tested.
- The second family of instruments use a different light source based on a selective emission of UV radiations. These lamps are selected to accelerate the decay processes induced by the sunlight using only the most energetic portion of its spectrum. The panels are mounted in sample holders about 30 mm from the fluorescent lights. Below, the panel is a tray containing water which is periodically heated up to provide condensation on the panel surface. Some QUV models are equipped with nozzles through which deionised water is sprayed onto the test panels. A typical test cycle would be 4-h UV light at 60 °C followed by 4-h condensation at 50 °C.

There are several standard methods based on particular weathering cycles using different combinations of the three phases: irradiation, water spray and water condensation. Among them are included the following standards:

- *ISO 11507 – Paints and varnishes – Exposure of coatings to artificial weathering – Exposure to fluorescent UV and water*
- *ISO 11341 – Paints and varnishes – Artificial weathering and exposure to artificial radiation – Exposure to filtered xenon arc radiation*
- *ASTM G154 Standard practice for operating fluorescent light apparatus for UV exposure of non-metallic materials*
- *ASTM G155 Standard practice for operating xenon arc apparatus for exposure of non-metallic materials*

An argument subject to frequent discussions is the comparison and degree of correlation between natural and artificial weathering tests.

This is a difficult problem since natural weathering does not correlate with itself from year to year. A way out of this difficulty is to develop a proper dosage model, for example, based on irradiance, time of wetness and temperature. Response to artificial weathering is system specific and the initial rank correlation with natural weathering must be corrected. Some people find this counter-intuitive. What is seems reasonable to mention is the possibility to rank in the same order a series of sample tested with the two systems but only when the coating composition is similar. The present accelerating methods must be considered as comparison methods rather than as absolute predictors of performance.

EN 927-6 specifies a method, assessment and evaluation for determining the resistance of wood coatings to artificial weathering in a QUV tester. A special spraying dry cycle was developed for this method [5].

6.2.2. Water permeability

Water absorption, in the form of rain, snow, hoarfrost or vapour can lead to different effects on wood. The first is the dimensional variation in which consequences are related to the final use of the product.

Functionality can be affected for certain elements as window frames which dimensional stability is fundamental for their proper use. Other types of products are less sensitive to possible dimensional variations (fences, garden furniture, etc.). This subject is considered by the EN 927-1 standard.

Dimensional movement is also a major contributor to coating failure on exterior exposure since it imposes a strain on the coating. Moisture trapped beneath a coating can cause blistering. Another consequence related to water absorption is the possible moulds growth or decay. This phenomenon can take place when humidity exceeds values around the 20%.

One of the roles of coatings for exterior use is that of limiting or modifying the rate of water (or) humidity absorption. The effect on performance is, however, complex and interacts with other properties including adhesion and mechanical properties [6]. One of the complications with permeability testing is the number of units or procedures for normalising the results. For some applications desorption, or the absorption/desorption ratio are the critical parameters.

The requirement of the coating treatment towards water exclusion depends on the final use of the product considered. European standard EN 927-5 Paints and varnishes – Permeability to liquid water describes a test method evaluating the permeability of coated wood to liquid water (Fig. 8).

6.2.3. Resistance against biological deterioration

Because of the organic nature of the main constituents, wood can be subject to the aggression of living organisms such as algae, fungi, bacteria or insects (see Chapter 2).

Figure 8   Permeability to liquid water according to EN 927-5 (picture from CATAS spa).

Test methods are divided between those that address decay of the wood itself, and those which address coatings which may be on the surface of wood. The European standardisation committee on wood protection has comprehensively treated the former subject and a number of standards have been produced:

- *EN 335-1:2006; Durability of wood and wood-based products – Definition of use classes – Part 1: General*
- *EN 335-2:2006; Durability of wood and wood-based products – Definition of use classes – Part 2: Application to solid wood*
- *EN 335-3:1995; Durability of wood and wood-based products – Definition of hazard classes of biological attack – Part 3: Application to wood-based panels*
- *EN 599-1:1996; Durability of wood and wood-based products – Performance of preventive wood preservatives as determined by biological tests – Part 1: Specification according to hazard class*
- *EN 599-2:1995; Durability of wood and wood-based products – Performance of preventive wood preservatives as determined by biological tests – Part 2: Classification and labelling*
- *EN 460:1994; Durability of wood and wood-based products – Natural durability of solid wood – Guide to the durability requirements for wood to be used in hazard classes*

The basic procedure for using EU Standards is to:

- Identify service environment and define hazard class using EN 335-1 and EN 335-2
- Determine natural durability and treatability of chosen species using EN 350-2

- Refer to hazard class and specify preservative according to EN 599-2
- There is no need to refer to method of treatment specifications; any process that achieves the desired result is permitted.
- Refer to EN 351-1 to define penetration and retention requirements

In the UK 'process' standards were traditionally used (rather than the so-called 'results' standards), and in accordance with EU law BS 5589 and BS 5268 have been withdrawn. The UK now has a national guidance standard *BS 8417:2003 Preservation of timber – Recommendations*.

CEN/TC 38 – Durability of wood and wood-based products is responsible for an important test for wood coatings: *EN 152 Test methods for wood preservatives – Laboratory method for determining the preventive effectiveness of a preservative treatment against blue stain in service – Parts 1 and 2*. This standard lays down a method for determining the effectiveness of a preparation applied by brushing or similar superficial treatment (e.g. spraying, spraying tunnel or dipping) resulting in an equivalent retention of product in preventing the development of blue stain fungi in wood in service. It is also applicable where a priming paint is used in conjunction with the preservative system.

6.3. Coatings for interior use

The degradation factors affecting coated surfaces in interior environments can be considered as consequences of the usual actions (continuous or accidental) carried out by the people during their daily life inside houses, offices or other living spaces. They can be either accidental events such as the falling of an object, or continuous stressing factors like surface cleaning, scuffing on floorings, etc. There are several test methods simulating the degradation effects affecting coated surfaces during the real use. They could be broadly sub-divided into four groups relating to the challenge that must be resisted:

(1) Mechanical stresses (impact, scratch, abrasion)
(2) Physical stresses (light, climatic variations)
(3) Wet and dry heat
(4) Chemical interactions (contact with liquid substances, e.g. in accidental spillage)

Very often, the various national or international standardisation committees define test methods for coated surfaces by considering the final use of a certain product, but not considering its composition. Moreover, performance requirements are generally not directly included in the standards. The 'philosophy' of some standardisation committees is to prepare specific instruments (technical test methods) but leaving the market free to define specific requirements.

In the case of floorings, the situation is different. The European committee TC 138 has defined some specific tests for parquet floorings and another technical committee (TC 112) has prepared a different standard (EN 14354) the title of which is *Wood-based panels – Wood veneer floor covering*. This document specifies test methods and requirements for wood veneer floor coverings for internal use. The European standard *EN 13696 Wood flooring (including parquet)* is a test method to determine elasticity and resistance to wear.

Within the furniture industry, the same or similar tests are applied to all types of finishes including plastic laminates and foils, as well as liquid coatings. It is the finish substrate combination that is tested, not the finish in isolation.

6.3.1. Mechanical stresses

6.3.1.1. Scratch resistance   Scratch resistance is not uniquely defined, and is difficult to measure. Different terminologies are used, ranging from marring for optical defects towards friction, wear, erosion, abrasion and cracking for mechanical defects. Scratch resistance can be evaluated in terms of gloss reduction, haze, magnitude of deformation, resistance towards deformation, weight loss or colour change of the substrate (Fig. 9).

A scratch can arise from of elastic, ductile and brittle components.

Scratch resistance is defined by *ISO 4586* as the minimum load applied to a needle of defined geometry that produces a continuous mark, visible to the naked eye, at the surface of the coating. ISO 1518 is used to assess the resistance of a dry film of paint, varnish or related product to penetration by scratching with a needle. The equipment consists of a horizontal motorised stage on which a coated panel is mounted. A weighted needle rests on the coated surface and forms part of an electric circuit which includes a meter. This meter deflects if the needle penetrates the coating and comes into contact with the metallic substrate. During the test the coated panel is driven under the needle and the meter is observed in order to see if the coating is penetrated through to the substrate. After the test, any scratch formed on the coating is examined visually in order to

Figure 9   Scratches on a table surface.

assess the extent and nature of the damage. The equipment can be used to determine the minimum load on the needle that will cause penetration through to the substrate. Alternatively it can be used to establish whether a specified needle loading will cause penetration.

Another test carried out with needles having of defined composition and shape able to produce a scratch onto the specimen surface is *ISO/DIS 4211-5 Furniture, test on surfaces – Assessment of surface resistance to scratching*.

Another similar method for furniture surfaces (CEN/TS 15186) is under development by CEN/TC207/WG7.

6.3.1.2. Abrasion resistance   Abrasion can be defined as the effect of a scraping or rubbing exerted by natural or artificial means onto the surface. The result depends on the ability of the surface to maintain the pattern, the colour or the original aspect under an abrasive action.

ISO 7784-2 describes a method which is intended for determining the resistance to abrasion of a dry film of paint, varnish or related product using rotating abrasive rubber wheel. The dried film is abraded, under specified conditions, with abrasive rubber wheels which are attached to an abrasion testing machine known as a Taber abrader. The wheels are loaded with specified weights and the test panels rotate under them for a specified number of cycles. A similar test for furniture surfaces (CEN/TS 15185) is under development by CEN/TC207/WG7. The resistance to abrasion is defined as either:

- the loss in mass after a specified number of abrasion cycles or
- the number of cycles required to abrade down to an underlying coating or to the substrate

For *ISO 7784-1*, the same test apparatus is used but plain rubber wheels with abrasive paper wrapped round them are substituted for the rubber wheels with embedded abrasive. Other systems are based on the falling of an abrasive powder (sand) onto the surface under investigation. The results of the abrasion resistance tests depend either on the mechanical properties of the coating and, for certain standards, also on the film thickness.

6.3.1.3. Resistance to impact   Impact tests evaluate the effects of accidental contact damage that may occur during use. They are usually carried out by the direct evaluation of the effect of a falling object, sphere or dart of specified shape and hardness, onto the tested surface form different heights. *ISO 6272-1* is used to assess the resistance of a dry film of paint, varnish or related product to cracking or disbondment from a substrate when it is subjected to deformation by a falling weight dropped under standard conditions. The result is based on the degree of damage assessed by reference to the number of cracks produced.

During impact on coated surfaces, the deformation of the substrate can be relevant, significantly affecting the final results. It is of importance that the coating film is flexible enough to withstand, to a certain extent, the stretching at the fringes of the deformation without any cracking or cleavage. Apart from the alteration in the surface appearance, the possible production of cracks enables dirt and moisture to penetrate into the substrate leading to different defects.

An important practical issue associated with impact testing is the typical inhomogeneity of wooden substrates. This may have a decisive influence on the repeatability and reproducibility of test results. A test method referred to furniture surfaces is described by the standard *ISO 4211-4 Assessment of surface resistance to impact*.

6.3.2. Physical stresses

6.3.2.1. Resistance to heat (humid and dry)   The resistance to heat is carried out to evaluate the effect produced by the contact of the surface under investigation with a hot object. It is a simulation of possible common actions happening frequently in a domestic contest where the leaning of a hot object (e.g. a cup of coffee) on to a furniture surface (table) is a very common.

There are also special methods interposing a wet textile between the two surfaces. Such type of tests called 'resistance to wet heat' is usually more severe as the effect of hot water (or vapour) may lead to discoloration or halo effects. The existing test methods regarding furniture surfaces are the following:

- *EN 12721 Assessment of surface resistance to wet heat*
- *EN 12722 Assessment of surface resistance to dry heat*
- *ISO 4211-2 Assessment of surface resistance to wet heat*
- *ISO 4211-3 Assessment of surface resistance to dry heat*

6.3.3. Resistance to climatic variations

Climatic variations can be the origin of different defects in coated surfaces. One such problem is the dimensional movement of wood in consequence of humidity variations. Also, expansion and contraction of coating films occurs in response to temperature fluctuations. Consequently, climatic variations can be responsible for the origin of possible tensions between the two materials becoming critical especially in the case of inadequate extensibility of the coating films (Fig. 10).

American standard *ASTM D1211 Standard test method for temperature-change resistance of clear nitrocellulose lacquer films applied to wood* describes a test method based on the execution of conditioning cycles from −20 °C up to 50 °C.

ISO 6270-1:1998 is a general test method for resistance to humidity − *Part 1: Continuous condensation*. In addition, there are several national test methods based on cyclic tests including, sometimes, also the control of humidity.

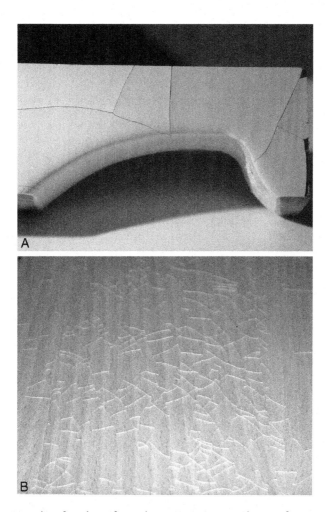

Figure 10   Examples of cracking after a climatic variations test (picture from CATAS spa).

6.3.4. Resistance to light

Light resistance is defined as the ability of a coated surface to maintain unaltered its aspect under the action of glass-filtered sunlight. Natural substances imparting colour to wood, or the colorants added to a coating product, can be altered in consequence of the exposure to light producing colour variations usually perceived as 'appearance defects' by the end users. Also, the colour of organic coating films, depending on their composition (e.g. polyurethane coatings based on aromatic isocyanates), can be altered by the effect of photo-chemical processes induced by light.

In the case of opaque paints, colour change induced by light can be considered as deriving from only from the possible photo-degradation and other chemical changes of the coating components themselves (binder or colouring substances). However, the discoloration of clear coatings is more complex depending on additional factors including:

- Colour change of wood substrate itself
- Degree of screening by the coating film towards the substrate

The composition of wood is different among the various species and a wood is an inhomogeneous material. Inevitably colour variations caused by light can differ considerably species to species but also between adjacent areas of the same wood element. The second parameter mentioned is related to the ability of a coating film to absorb the radiations causing such colour change, limiting their effect onto the substrate. These radiations are usually the most energetic of the spectral range (UV) and so the addition of specific additives (UV absorbers and hindered amine light stabilisers HALS) is the strategy usually adopted by the coating manufacturers to improve such protection capacity for clear coatings.

However, UV radiation in interior environments is relatively limited due to the filtering effect of window glass. Very often, it is the visible radiation which is causing colour changes. The use of UV absorbers is then not a definite solution for such problem and specific Lignin stabiliser may be considered [7].

Tests to evaluate the light resistance are carried out by particular systems equipped with xenon lamps. There are also special filters reducing the IR and UV portion in order to make the emission spectrum similar to the glass-filtered sunlight. At the end of the exposure period, duration of which depends on the irradiation specifications, the colour change of the exposed sample is evaluated in comparison with a non-exposed reference.

Standard *ISO 11341 – Paints and varnishes – Artificial weathering and exposure to artificial radiation – Exposure to filtered xenon arc radiation* describes the general principles for the artificial weathering of coating system both for interior and for exterior use. In addition to this standard, EN 15187 specifies a test method for furniture surfaces. Colour changes on the surface of coated wood can also be caused by extractives, and resin from knots. CEN TC139/WG2 is currently developing test methods to quantify these effects [prEN 927-7].

6.3.5. Chemical interactions

6.3.5.1. Resistance to liquid substances   The chemical resistance of a coated surface can be defined as its ability to maintain unaltered in its appearance when subject to the action of certain chemical agents under specific conditions (Fig. 11).

ISO 2812-1 Paint and varnishes – Determination of resistance to liquids refers specifically to the resistance of applied coatings towards liquid

Figure 11 Coffee drops on a table surface.

substances commonly used in interior environments (cleaning products, drinks, ink, oil, disinfectants, etc.).

In addition to this standard, *EN 12720 Resistance to cold liquids* (similar to *ISO 4211 Assessment of surface resistance to cold liquids*) specifies a test method for furniture surfaces.

These test methods generally specify to apply each substance by means of adsorbent pads imbued with the liquid and covered by a watch glass. After the prescribed period, the effects are evaluated.

There is also a similar test method standardised for floorings *EN 13442, Wood and parquet flooring and wood panelling and cladding – Determination of the resistance to chemical agents.*

REFERENCES

[1] Orchard, S. E. (1994). Flow of paint coatings: A hydrodynamic analysis. *Prog. Org. Coatings* **23**(4), 341–350.

[2] Hunt, R. W. G. (1991). "Measuring Colour", 313 pp. Ellis Horwood Ltd, Chichester (ISBN 0-13-567686-X).

[3] Bulian, F. (2006). La protezione del legno dall'acqua. *Pitture Vernici Eur. Coatings* **82**(10).

[4] Martin, J. W., *et al.* (2006). Making a linkage between field and laboratory exposure results via a reliability-based methodology. *In* "Proceedings of the Federation of Societies for Coating, Technology, '2006 FutureCoat' (ICE 2006) Conference", New Orleans, Paper 1.

[5] Podgorski, L., Arnold, M., and Hora, G. (2002). Reliable artificial weathering test for wood coatings. *In* "Proceedings of the PRA 3rd International Wood Coatings Congress 'Wood Coatings: Foundations for the Future'", 15 pp., The Hague, Paper 13.

[6] Graystone, J. A. (2001). Moisture transport through wood coatings: The unanswered questions. *Surface Coatings Int. Part B: Coatings Trans.* **84**(B3), 177–187.

[7] Schaller, C. (2008). New concepts for light stabilisation of coloured wood. *In* "Proceedings of the PRA 6th International Wood Coatings Congress", 14–15 October 2008, Amsterdam, Paper 10.

BIBLIOGRAPHY

Bulian, F. (2006). Il fenomeno del calo delle superfici verniciate. *Pitture & Vernici Eur. Coatings,* Milano **82**(10), 5–8.

Bulian, F. (2006). La protezione del legno dall'acqua. *Pitture Vernici Eur. Coatings* **82**(10).

Bulian, F. (2004). Test methods for interior coatings: A critical overview, atti COST E 18, Copenhagen.

Bulian, F., and Cianetti, E. (2002). Prodotti vernicianti per i mobili – Certificazione di prodotto. *De Qualitate anno XI* **5.**

Bulian, F., and Tiberio, M. (2002). "Distretto della Sedia: Gli aspetti ambientali e le migliori tecnologie", vols. 1 and 2, CATAS.

Frova, C. A. (2000). Luce colore e visione, BUR, Milano.

Koleske, J V. "Paint and Coating Testing Manual", ASTM PCN 28-017095-14.

Mc Donald, R. (1987), Colour physics for industry, Society of Dyers and Colourist.

CHAPTER 7

Market Needs and End Uses (1) – Architectural (Decorative) Wood Coatings

1. INTRODUCTION

Previous chapters have described the building blocks (raw materials) from which wood coatings are formulated and some of the advantages and constraints that are associated with each delivery technology (water-borne, solvent-borne, radiation curing and powder). Important wet and dry film properties have also been introduced. The role of the formulating chemist is to select components that will deliver the balance of properties and appearance required for a specific market sector. Superficially, there

Wood Coatings: Theory and Practice
DOI: 10.1016/B978-0-444-52840-7.00007-2

is considerable overlap between the decorative and industrial sectors as indicated in Table 1. However, there are very substantial operational and performance requirements such that the technological choices are usually different.

Table 1 Decorative and industrial wood coating sectors

	Architectural	Industrial
Interior end uses		
Internal doors Interior window surfaces Panelling and trim	Emphasis on maintenance and redecoration. Air drying technology only	Industrial coating processes. Several competing technologies
Flooring	In situ DIY and specialist maintenance sector	Automated fast cure factory processes. Laminates and parquet
Furniture Knockdown (flat stock) Assembled (3D) Profiles and mouldings	Relatively minor niche sector	Major industrial category; primarily automated. Competing technologies. Several sub-sectors. Wood and wood derivatives
Exterior end uses		
Joinery and Windows Doors Cladding Shutters	Usually maintenance, DIY and Trade. Range of different exterior air drying products	Factory application, different products and operations from furniture
Decking Fencing	Primarily stain products applied in situ	Not usually finished; preservative treatments
Garden furniture	Range of products (oils, stains, varnishes and some paints). Primarily DIY	Top end of market only, competition from metal substrates

1.1. Summary of key differences

(1) *Technology*. Architectural coatings are manually applied, and must dry and cure under ambient conditions. Thus, the main binder is typically a latex or alkyd. VOC of the products are limited in Europe by the solvent emissions directive (as noted in Chapter 1) and the role of solvents for interior products is increasingly curtailed. Industrial wood coatings have a wider gamut of technologies to choose from and processes are often wholly or partially automated. Volatile emissions are also curtailed, but factory conditions offer additional options such as abatement.

(2) *Maintenance or new work*. While it would not be true to say that the decorative sector *never* coats new work, it is certainly the case that there is much more emphasis on the maintenance of existing coatings. This creates a need for ancillary products such as stoppers and fillers. Even when coatings are removed (by paint strippers or hot air), the surface is not the same as new wood. In contrast, the Industrial sector is primarily concerned with new wood (or derived products such as particle boards and MDF). This gives greater opportunity to address the nature of the surface and the need to deal with cutting and sanding operations.

(3) *Interior or exterior*. Although the architectural sector does offer general purpose products, there is a growing tendency for exterior products to be formulated differently. Thus, exterior paints will be based on more flexible resins than interior ones, and varnishes and stains require additional protection from UV and mould growth. In the industrial markets product sectors are mostly defined by their function as interior or exterior, and formulated accordingly. Industrial joinery, especially windows require very durable products if they are to compete with alternative substrates such as UPVC and aluminium. Furniture is primarily for interior use though some garden furniture may be factory finished. However, a substantial amount of garden furniture is sold unfinished or stained, and subsequently finished using decorative products, this is often also true of decking and fencing.

(4) *Furniture sector*. The interior furniture sector is primarily an industrial operation with a very wide range of technologies the choice of which will depend on the desired appearance and whether coating is carried out on flat or assembled units. Within the decorative sector, this is more of a niche market including restoration.

(5) *Flooring sector*. The coating of wooden floors bridges both the industrial and DIY market sectors. In situ wooden floors may be restored with DIY products but there is also a semi-industrial operation for maintenance within the home, and in larger areas such as schools and sports halls using industrial products and processes. Real wood parquet and laminate flooring will also be coated on high speed factory lines prior to installation.

2. DECORATIVE COATINGS FOR EXTERIOR WOOD

Designing or selecting a wood coating requires a specification in terms of appearance and the projected end use. In the past, the main choices of appearance were described by the archetypal terms 'paint', 'varnish' and 'stain'. To draw a distinction with interior and decorative stains, products for exterior wood were usually designated as 'exterior stains', and in many countries as 'lasures'. As the market developed so did intermediate products, often with proprietary as well as generic names. Hence as was already described in Chapter 5, a European Standard was developed which allows for sixty different product appearance descriptions as a matrix of build (film thickness, four levels), gloss (five levels) and opacity (three levels) excluding colour. This is more than enough to cover the many products now available in the marketplace. It should be noted that the EN 927-1 classification refers to the total system which might comprise several coats of the same, or different, products (e.g. primer + two finishing coats).

In Chapter 5, it was also shown that the generic nature of coatings reflects their volume solids and pigment volume concentration (PVC) which are conveniently reflected in a tri-linear phase diagram (Fig. 1).

The formulating area covered by EN 927-1 is approximately indicated by the shaded area below; such that when the films dry (and lie along the pigment binder axis) they meet the film thickness criteria. Although the

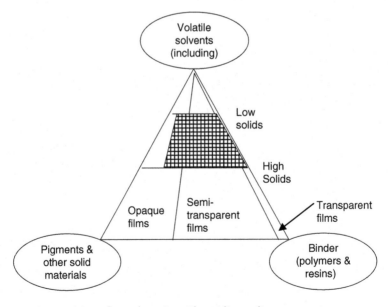

Figure 1   Composition of wood coatings. The tri-linear diagram.

dry film thickness is governed by the solids content, it will be modified by the spreading rate and number of coats. The transition from semi-transparent to opaque coatings will not be sharp and will depend on film thickness, and the hiding power of the pigmentation system. It is not possible to indicate the five gloss levels of EN 927-1 on the phase diagram since these will be composition specific. Generally, gloss is reduced at higher PVCs, and for interior decorative coatings a lower gloss, or sheen, is achieved through the use of extenders. This is less advisable for exterior coatings since the presence of extenders raises the modulus of elasticity, and reduces the extension to break. It is also less suitable for transparent coatings; alternatives are either to use wax, which also confers water repellency, or to use additional polymers with a refractive index different from that of the main film former.

In formulating a wood coating, it is therefore necessary to define the compositional parameters (especially volume solids and PVC) able to deliver the required generic product, for binders capable of meeting the requirements of specific end-use sectors.

In EN 927-1, end uses for exterior wood are categorised according to the degree of dimensional stability required. 'Stable' categories cover joinery, windows and doors; 'semi-stable' would cover many types of cladding while 'non-stable' includes most types of fencing (Table 2).

The term 'stability' is used here to indicate the extent to which the rate of wood movement in response to moisture should be modified by the coating. As noted in Section 6.2.2 of chapter 6, an important property related to substrate stability is water permeability. Other important film properties include extensibility, modulus of elasticity and adhesive performance, which must resist the inevitable movement of the substrate even with controlled permeability.

One consequence of a high coating permeability is a tendency of the wood to cracking and splitting. This is indicated by the test panels in Fig. 2.

Table 2 End-use category according to EN 927-1

End-use category	Permitted dimensional movement of wood	Typical examples of end-use categories
Non-stable	Free movements permitted	Overlapping cladding, fencing, garden sheds, decking
Semi-stable	Some movements permitted	Tongue and groove cladding, wooden houses and chalets, garden furniture
Stable	Minimum movements permitted	Joinery including windows and doors

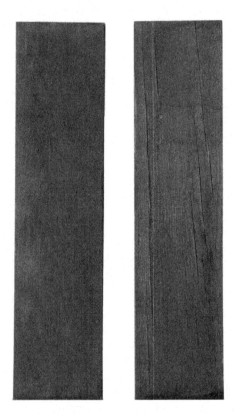

Figure 2   Test panels exposed outdoors (12 months).

The left-hand panel is formulated at a higher solids content, and consequently has a lower moisture permeability than the coating on the right.

After only 12-months weathering, the lower solids product is showing
considerable wood splitting. This would be unacceptable in joinery and
some cladding, but might not be regarded as a problem for rough sawn
timber or fencing.

Another consequence of unmodified moisture movement is that
the dimensional change can cause the jamming of windows within
their frame. Excessive movement also places a strain on gaskets and
sealants.

It should be stressed that there is no unique 'correct' solution to the
formulation or selection of a wood coating for any given application. The
desire for a high-build transparent coating, such as a yacht varnish,
carries the penalty of more frequent maintenance than an equivalent
opaque paint system. Lower solids stains are easy to apply and maintain,
but also require more frequent maintenance. In each case, the formulator

must come up with an optimum solution within the constraints of the broad generic type and other operational needs.

Some general remarks may be addressed to the three main generic classes as specified in the following paragraphs.

2.1. Exterior wood stains (lasures)

Archetypal stains are designed to penetrate, and will generally have a relatively low (<20%) volume solids content. Penetration into new wood can be studied by microscopy directly, or by additional labelling with a fluorescent agent. Radioactive labelling and radiography have also been used. Such studies show that there can be differential penetration between solvent, fungicide, polymer and pigment. Different molecular weight species will also show different depths of penetration. Penetration is usually deeper with solvent-borne compositions, whereas water causes swelling. However, penetration by alkyd emulsions is sometimes found to be closer to solvent-borne alkyd solutions than to water-borne acrylic dispersions. The physical form of the binder has a major effect and polymeric dispersions will not penetrate unless of very small particle size. Penetration by pigment is also constrained by size and only very finely ground pigments such as transparent iron oxide will differentially penetrate some wood cell structures.

Some penetration into the substrate is important for products including primers that are applied directly to bare or denatured wood. Penetration provides a degree of mechanical key but also improves the cohesive strength of the wood surface; this will improve the overall adhesive performance and reduce delamination of subsequent coats. Penetration also reduces the stress gradient transition between the coating and the substrate with an overall improvement of mechanical properties.

A distinction should be made between the functions of a wood stain and a wood preservative. The latter are designed to prevent decay and require special application methods (such as double vacuum treatment) to ensure deep penetration. Stains may contain fungicide to inhibit surface colonisation but normal brush application will not give sufficient penetration to ensure protection against decay. Stains and indeed all wood coatings on non-durable wood should be applied to preservative treated wood if there is a risk of high moisture content. It would be mandatory in the case of direct ground contact.

Penetrating stains can be formulated with little binder, often no more than is needed to stabilise the pigment. In solvent-borne products, oils and alkyds are often used, but care must be exercised when formulating at low vehicle solids as there can be problems of phase separation or oxidative gelling on long-term storage of the product; settlement can be another problem. Water-borne stains can be based on low volumes of aqueous dispersion polymer. Recent years have seen a proliferation of low-cost water-borne stains, especially for fencing and garden applications. These were often positioned (in marketing terms) as an alternative creosote (tar oil derivative) which is now banned, and have handling benefits with little or no effect on adjacent plant life. Such products are essentially decorative and have relatively little protective or preservative effect.

Low solids stains may be modified with silicone oils, waxes, etc., to give some, albeit temporary, water repellence but they exert virtually no control over water vapour uptake. Their use is thus best confined to situations where free movement of the substrate is acceptable. Typical applications might include fencing and some types of cladding, though the latter must be of a design that allows for movement. Good performance on cladding requires a sawn, rather than smooth planned surface. As a generalisation, it would be true to say that may wood coatings, even gloss paint, will perform better on a sawn surface, though naturally the appearance of gloss products will be affected.

In the case of joinery, especially softwood, there is a definite need for moisture control, otherwise constantly fluctuating movement can cause splitting, loosening of glazing and the seizure of opening lights. There is also a body of opinion which holds that too low a permeability will allow moisture to become trapped and increase the chance of decay. This line of thinking led to the promotion of some coatings as 'micro-porous' and other terms such as 'breathing' to imply a higher level of permeability than a traditional paint system. Strong evidence for a critical lower level of permeability has not been forthcoming and more weighting is currently given to design and wood quality factors.

Attention is also now focused on the distribution of water just below the coating surface and the role that the coating composition might have on this. There has been concern that surfactants at the wood surface below the coating might mimic the condition underneath wet bark in fallen forest timber. Decay organisms such as *Dacrymyces stillatus* may have evolved to exploit this niche.

Until more evidence is available a prudent compromise will be to control permeability for joinery between an upper and lower limit defined by the wood species and end use and to use a solvent-borne, rather than water-borne, penetrating priming coat (Table 3).

The present version of the performance standard EN 927-2 suggests (non-mandatory) permeability limits of 175 and 250 g m^{-2} for stable and semi-stable end uses using a test method prescribed in EN 927-5. There are no limits for non-stable applications.

The rate of moisture transmission in wood stains can be reduced by raising the volume solids, that is, moving to a higher solids stain, thus giving a thicker film at similar spreading rates (Table 4).

Table 3 Exterior stains (lasures): Summary of key features

Low solids	High solids
– <20% volume solids	– >20% volume solid
– Traditionally solvent-borne	– Both solvent- and water-borne
– Oil- and alkyd–based	– Alkyds, acrylic, 'hybrids'
– Transparent iron oxides	– 'Opaque' and transparent iron oxides
– Blue stain protection	– Blue stain protection
– Water repellents	– UV absorber
Key properties	
– Penetration	– Mechanical properties
– Colour retention	– Adhesion
– High permeability	– Medium permeability
	– Primer basecoat necessary
End use	
– Fencing	– Cladding
– Sawn cladding	– Joinery
– Priming coat for higher solids coatings	– Garden furniture

Table 4 Aqueous wood stain: Typical formulation range

Component (ingredient) raw material	Low-build stains (%) weight	High-build stains (%) weight
Binder solids[a]	10–20	25–45
Co-solvent	0–5	0–10
Pigments (e.g. trans iron oxides)	1–2	2–5
Additives (dispersant, biocide, thickener, etc.)	1–3	2–5
Water	60–80	30–60
Total solids	15–30	20–60

[a] Binders will typically be acrylic/acrylate dispersions, sometime as hybrid mixtures with an alkyd emulsion.

However, some care in formulation is needed. For some binders, raising the film thickness could change the failure mode from one of erosion to flaking, which negates one of the benefits of using a stain. Binders must be selected with a relatively low elastic modulus but high extensibility and adhesion. Because penetration of higher build stains is reduced, it is usually beneficial for the first coat to be formulated at lower volume solids.

Solvent-borne high-build stains can be successfully formulated using alkyds, bearing in mind the remarks made above. A number of proprietary alkyds have been developed specifically for this purpose. Water-borne transparent stains for joinery proved more difficult to formulate. There are three major problems to overcome: firstly, the much higher permeability creates moisture control problems; secondly, fungal attack is more common; and finally, transparency of acrylic and vinyl polymers to UV light makes it more difficult to prevent photo-degradation of the surface. It has been found that many of the vinyl and acrylic latexes which perform well in paints (i.e. including opaque stains) are not suitable for transparent coatings. Attempts have been made to overcome this problem by incorporating components capable of absorbing UV light and by the use of hybrid systems combining acrylic dispersions with alkyd emulsions. Good results have also been reported with polyurethane dispersions. This is an area of continued development.

The pigmentation of wood stains is relatively straightforward; the demand for 'natural' wood shades makes red and yellow iron oxides particularly useful. The best colour, brightness and transparency are achieved with synthetic iron oxides which achieve full transparency when ground to colloidal dimensions. However, these are expensive materials and the standard grades require long dispersion times. Their high surface area is quite active and some care in selection of dispersant must be exercised to achieve long-term stability. Specialist dispersant suppliers can provide recommendations and supporting evidence for solvent- and water-borne systems. When fully dispersed these grades confer exceptionally good UV screening; however, colours become progressively darker at higher concentrations, and for pale transparent colours additional protection may be required. This need was first established for the clear coats used over metallic basecoats in automobile coatings. A combination of UV absorber and hindered amine light stabiliser (HALS) proved an effective solution, and this technology has been extended to wood coatings (see Section 2.2).

The choice of fungicides for wood stains is considerable but their compatibility and long-term storage must always be checked very carefully. For fencing and ground contact applications, copper and zinc soaps perform well, and have relatively low mammalian toxicity; however, concerns have been expressed about the role of copper in aquatic environments. Stains and other transparent coatings for wood should include fungicides which are specific against blue stain. Water-borne stains, and indeed water-borne coatings in general are more vulnerable to surface mould and solvent-borne (Table 5).

This arises because they are generally more permeable and may contain more available nutrients; however, the effect is formulation specific. In selecting one of the many fungicides now offered for wood coatings, it is extremely important to assess the longer-term storage stability; some

Table 5 Dry film biocides (examples only)

Biocide	Notes
Carbendazim (methyl benzimidazole-3-carbamate)	Limited activity spectrum, effective for blue stain
OIT (2-n-octyl-4-isothiazolin-3-one)	Rather soluble (dichloro-OIT less so)
Chlorthalonil (1,3-dicyano-2,4,5, 6-tetrachlorobenzene)	Good activity spectrum but tends to hydrolyse in water (and yellow)
Diuron (N'-3,4-dichlorophenyl-N, N-dimethylurea)	No anti-fungal activity, used in combination with fungicide
IPBC (3-iodopropargyl-N-butylcarbamate)	

have shown seeding or discoloration effects, which may take several months to appear.

The term 'wood stain' has been used generically in the above discussion; within the broad 'woodcare' market products will be positioned in specific sectors such as 'decking' or 'fencing' with many variants. Formulations will vary according to the specific sector; for example, decking finishes often contain proportionally more water-repellent additives. Some care may be needed in re-coating solvent-borne decking finishes with a water-borne one as inter-coat adhesion can be poor.

2.2. Varnishes and other clear coats

A 'clear coat' is defined as one which when applied to a substrate forms a solid transparent film having protective, decorative or specific technical properties. It is more general than the term 'varnish' which refers to a clear coat drying by oxidation; however, in practice the terms may be used interchangeably. Although archetypal varnishes are glossy there has always been a demand for mid-sheen (satin) and matt variants, particularly indoors. Some exterior products also have a lower gloss and to some extent the technology has evolved to overlap with high-build wood stains.

Traditional exterior varnish usage has declined steadily in favour of wood stains, due to a high incidence of flaking and discoloration with consequent increased maintenance costs. However, there remains a demand for good quality varnish to set off the aesthetic qualities of wood. Varnishes also have better wear characteristics in high traffic areas such as doors and door frames.

Problems with conventional oil and alkyd varnishes arise from the fact that they are high-build films of low permeability, becoming brittle with

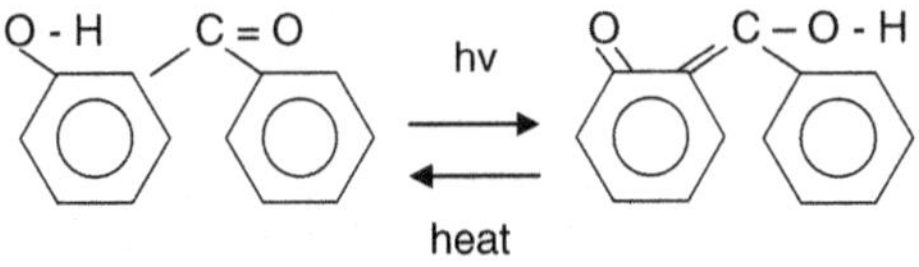

Figure 3 Dissipation of UV light (as heat by tautomerism).

age. The problem is made worse by their transparency which allows photo-degradation at the wood varnish interface. Traditionally, good quality varnishes were based on tung and linseed oils, modified with phenol formaldehyde resins. Straight alkyds are generally less effective. It is probable that the good performance associated with tung–phenolics is associated with their UV absorbing characteristics and improving this aspect of performance has received considerable attention, notwithstanding the fact that visible light is also detrimental to the substrate.

Improvements in performance have been claimed with several types of UV absorber (Fig. 3), including the benzophenone groups and HALSs. Benefits have also been noted with inorganic materials, including nano-size zinc oxide and titanium dioxide. Yellow transparent iron oxide, as used in wood stain, also confers a marked improvement, though clearly the colour will darken. In practice, this may be no worse than the yellowing which occurs with many varnishes on weathering. The amount ant type of UV absorber will depend on the opacity of the coating [1] (Table 6).

It has not been customary to include fungicide in varnishes but logically they should also contain agents to inhibit blue stain and other fungi. Many fungicides are themselves degraded by UV radiation so the benefits are seldom realised unless the varnish formulation does screen out UV. Adhesive performance, and hence resistance to flaking, can be improved by thinning the coat first applied to the wooden surface or by substituting a low solids wood stain. Such an approach further blurs the distinctions that can be drawn between high solids stains and coloured varnish.

Demand for exterior varnishes is usually in the full gloss form, but satin and matt are also popular for interior use. Settlement of matting agents requires careful formulation, for some can interact with structuring agents. Traditionally, the majority of external varnishes remained solvent-based, reflecting the difficulty of achieving good flow and high gloss with aqueous dispersion binders. Externally, there is also the problem of durability already mentioned in the context of stains. However, aqueous acrylic and styrene–acrylic latexes with in-built UV protection can provide the basis of an exterior varnish. For interior use, there has been a steady growth in the adoption of polyurethane dispersions; this technology is also showing promise for external coatings. Needless to say, there is a wider choice of binder systems for industrial wood coating.

Table 6 UV protection: Surface and bulk effects

UV absorbers. They typically absorb in 290–380 nm range but newer types extend to 400 nm (more suitable for wood):
- Absorption depends on path length, therefore more effective in bulk than at the surface
- Some UV absorbers (light stabilisers) also act to quench photo-excited polymers
- The effect is polymer specific

Anti-oxidant. They can supplement protection:
- By decomposing peroxides
- Complexing with metal ions
- Interfering with chain propagation

HALS. They can act as chain-breaking anti-oxidants:
- Whereas UV absorbers reduce the rate of radical generation, HALS reduces the subsequent rate of oxidative degradation
- HALS is effective in surface and not just bulk
- Therefore together HALS + UV absorber is synergistic
- May be deactivated by chemisorption onto other components, especially pigments.

Lignostab® 1198. It is a proprietary photo-stabiliser developed specifically for wood pre-treatments (primers):
- It is high effective for the colour stabilisation of natural, tinted or stained wood
- It improves the durability of wood substrates coated with clear and transparent pigmented finishes

---

2.3. Paint and paint systems for wood

As noted earlier, the term 'paint' may be used as a generic umbrella term to cover opaque pigmented coatings, regardless of sheen, build or system. In many countries, there has been a tradition of using high-build full gloss systems on woodwork, especially outside. During the past decade, the supremacy of gloss systems has been challenged first by wood stains, and more recently by paints with a sheen level lower than full gloss. Despite these changes, the demand for a full gloss system remains high, especially for redecoration, and this will remain an important market sector in France, Germany and the UK for the immediate future. In other parts of the world, including North America and Australia, the move away from high gloss exterior systems has gone further. This situation can be explained in a number of ways, but it is likely that different construction methods and the greater use of wooden siding (cladding) would show up potential defects of the traditional system (such as flaking) on a larger scale, thus paving the way for greater use of more permeable off-gloss water-borne systems.

Traditionally, high gloss paints were part of a three-product system comprising primer, undercoat and gloss. Many of the new paints are two-product, or one-product multi-coat systems. Before embarking on the design of a new coating, some consideration must be given to the merits of a system as opposed to a single-product approach.

An obvious advantage of single products is simplicity, to both the user and the stockist. Against this, it must be recognised that a coating has many different functions to perform which might include sealing, adhesion promotion and filling, as well as protection and appearance attributes. Combining these onto one product is likely to entail compromise which may or may not be acceptable to the final customer.

If two or three coats of an alkyd topcoat are applied directly to bare wood the durability can be remarkably good, better in many cases than a traditional three part system. What then has been lost? Answers to this question include build (film thickness – affecting appearance), speed of re-coating and flexibility in redecoration. What weighting is given to these factors varies between individuals and user groups. Many of the newer products will inevitably be used for redecoration of existing coatings rather than new wood. They may have good durability but show disadvantages for the practical problems encountered in both trade and retail use. Whatever the detail of the final chosen system, formulators must consider how the needs of priming, filling and finishing are to be met.

2.3.1. Wood primers

An important function of wood primers is to provide adequate bonding between the substrate and subsequent finishing coats, especially under damp conditions. Primers should be able to seal end grain while resisting hydrolytic breakdown over long periods. Some building practices exposes building components for variable periods without the protection of a full system; hence, primers should have sufficient intrinsic weather resistance to provide a sound base for subsequent re-coating.

In the past, the archetypal wood primer was based on linseed oil and white lead. Lead carbonate has the useful property of forming with linseed oil, fatty acid soaps which have good wetting properties and yield tough, flexible films. White lead was often admixed with red lead (lead tetroxide) giving a characteristic pink wood primer. Recognition of the cumulative toxicity of lead has led to the withdrawal of lead-containing products and the development of products without lead pigment. In the UK, a derogation clause allows the continued use of this type of primer only on heritage buildings. This is partly driven by a desire to use traditional products on such buildings, but it also reflects the high esteem in which the performance of lead-based primers is held and many would regard them as a benchmark for performance.

The immediate replacements for traditional lead primers were based on oil, oleoresinous or alkyd binders with conventional pigmentation. To maintain flexibility, the PVC needed to be relatively low (typically 30–40%). Early low lead primers were often formulated at a much higher level (to reduce blocking) and this, combined with fast-drying inflexible alkyds, resulted in poor weathering performance, a factor contributing to dissatisfaction with conventional solvent-borne systems in general. For improved solvent-borne products, a key requirement is to use an extensible binder. Weathering trials have shown advantages in using a combination of some free drying oil, as well as that which is chemically bound to the polyester backbone of an alkyd resin. This may reflect the ability of free oil to penetrate the wood more effectively. It used to be said that the oil was 'feeding the wood'. Early attempts to formulate water-borne primers for solvent-borne topcoats were disastrous with premature flaking of the topcoat occurring. The problem was partly the result of a mismatch between the two sets of mechanical properties, but very much exacerbated by poor inter-coat adhesion. As acrylic latex technology developed, it was found possible to overcome this problem by incorporating specific adhesion promoting groups into the polymer. This is an area of proprietary know-how, and is also used to improve the performance of interior water-based paints that are used in damp conditions subject to condensation such as kitchens and bathrooms.

The EN 927-2 performance specification for exterior wood coatings does not set out to test or specify individual components of a paint system. It is assumed (e.g. in the mandatory weathering test EN 927-3) that the whole system will be tested. However in the UK, there is a performance standard for primers alone (BS 7956:2000 – Specification for primers for woodwork) which includes a weathering and re-coating test. This reflects the possibility of providing primed windows which are later finished in situ. Better performance is achieved using full factory-finished systems. Another feature of the BS 7956 is a blister-box test. This has shown good ability to predict adhesive performance under wet conditions of solvent-borne topcoats. Generally, systems which pass the blister-box test show good exterior durability. The test will show up, for example, the difference between interior and exterior grade latex binders. However, the test is less relevant when water-borne topcoats are used.

2.3.2. Aluminium sealers and wood primers

Primers containing aluminium flake offer good barrier properties to both water and certain types of staining. They are useful for sealing end grain and against resin exudation, creosote, bitumen and coloured preservatives. Barrier properties are dependent on the amount of flake pigment used. Established vehicles for this type of product include phenolic-modified resins. In the UK, typical formulations are specified in BS 4756:1998. Solvent-borne aluminium primers are not advisable on large areas where high movement may be expected. A high incidence of flaking has been found on softwoods, reflecting poor inter-coat adhesion. Performance is usually better on hardwoods.

Water-based aluminium primers were slower to develop due to the propensity of aluminium pigments to form hydrogen gas in the presence of water. However, this problem was solved (with passivated aluminium pigment grades) originally developed for the car industry.

2.3.3. Preservative primers

This term is sometimes used to describe an un-pigmented fungicide-containing binder. Such products are designed for maximum penetration at the expense of any contribution to film build. It is thus easier to achieve penetration of fungicidal components, especially into the vulnerable end grain.

Inclusion of a suitable low viscosity binder enables the product to seal end grain to a greater depth than can be achieved with a pigmented product, though more than one application may be necessary. The penetrating nature of these products can help stabilise a denatured surface and improve subsequent coating performance. For completeness, it should be noted that the term 'preservative' has also been used as an adjective to describe any products including stains and primers which contain fungicide. Manufactures of such products must consider labelling implications in the light of the Biocidal Products Directive.

2.3.4. Dual-purpose stain primers (stain basecoats)

Because there is a possibility that new joinery may be finished with either stain or paint, it may be an advantage to joinery manufacturers to prime with a primer that can be over-coated with either type of product. This has created an apparent niche for dual-purpose primers. Their appearance is inevitably dictated by the needs of the semi-transparent stains and such products tend to be closer to stain primers than traditional wood primers. In consequence, they offer much less temporary protection than a wood primer, and this has caused problems. There is also a danger that this approach obscures from users the fact that stains and paints are not always fully interchangeable for a given construction. Highly permeable products may require a better quality of joinery wood if excess movement is to be avoided and will require alternative glazing and non-ferrous fixings.

2.3.5. Knot and tannin staining resistance

A function often required from a priming coat is the prevention of discoloration. The presence of various extractives in wood was introduced in Chapter 2. Knots are associated with resin but also contained coloured materials. Tannins are water-soluble stains present in the heartwood of certain woods such as western red cedar but also including woods like oak, jarrah, meranti, merbau and many others. Because tannins (gallic acid-3,4,5-trihydroxybenzoic acid) are water-soluble, they present a particular problem for white water-borne coatings which is not yet entirely resolved.

Traditionally under opaque coatings, knots were sealed with a solution of shellac carried in methylated spirit, or with aluminium pigmented paint. Although this was not completely successful, it usually sufficed for solvent-borne products. Gently heating knots and wiping off any excess resin may improve performance, as does multiple spot priming.

Staining from knots and tannin is different and difficult to test; prEN 927-7 is a draft test method for knot staining but it has not yet proved possible to develop a repeatable and reproducible European test method for tannin staining, though work is in progress. The staining of white topcoats is seen as a major commercial problem and has engendered considerable research activity. Many specific compounds have been patented and are used in proprietary products; they include:

- Ion exchange resins
- Organo-silane compounds (with acid-rich polymers)
- Zirconyl acetate
- Ammonium zirconium carbonate
- Titanium chelates
- Transition metal complexing agents
- Magnesium hydroxide
- Zinc oxide
- Polymers with specific functional groups

Other formulation components can also influence the tannin-blocking capability of wood primers. Charge stabilised dispersants have been reported as more effective than steric stabilisers with HEUR (hydrophobic ethoxylated urethane) thickeners also giving better non-bleed characteristics than other thickener types [2].

2.3.6. Undercoats

Undercoats can play a useful role in contributing build, opacity and colour to the traditional paint system. They also improve adhesive performance when old gloss is repainted and help provide cover on sharp edges. Achieving a high gloss appearance when renovating weathered gloss paint is more difficult without an undercoat to fill damaged areas and provide a contrast between coats. To fulfil these functions, undercoats are normally heavily filled and habitually have the highest PVC of any exterior coating. Moreover to aid sanding, they are usually formulated on a brittle binder. As a consequence, they lack extensibility and from the point of view of durability, are the weak link in a paint system. Replacing undercoat by an extra coat of gloss significantly improves the durability of many paint systems, though this would be regarded as less practical by some decorators. The fact that some undercoats are inextensible does not invalidate their usefulness, but exterior undercoats for wood should be formulated differently from those for interior or general purpose. In particular, they must have greater extensibility which can be achieved by reducing the PVC or by using a more flexible binder. This will, of course, increase sheen and makes sanding more difficult since the abrasive paper quickly becomes clogged. Requirements with respect to adhesion are less stringent if a good primer is used, and there are many potential formulating routes via water-borne or solvent-borne technology. In general, traditional undercoats are more suitable for interior surfaces than exterior.

2.3.7. Finishes (topcoats)

Full gloss solvent-borne gloss finishes have remained an important feature of many European markets, especially in maintenance painting. When used with primers and undercoats (optional) which have been correctly formulated for use on wood, they can give performance with a typical maintenance period of 5–7 and exceptionally 10 years. There are sufficient differences between interior and exterior conditions to make separate formulations worthwhile. In particular, exterior gloss finishes for wood will benefit from increased flexibility and the presence of fungicide to inhibit blue stain and other surface moulds. Gaining the optimum balance between longer-term durability and initial drying properties requires fine tuning. Alkyds (oil- or fatty acid-modified polyesters) lend themselves to almost limitless modifications, which alter mechanical properties, permeability, adhesion, etc., and have a marked effect on durability. Over the 4–5 decades that alkyd resins have been in commercial use, there have been many attempts to relate their structure and chemistry to properties such as durability, but due to the complexity of the situation, clear and unequivocal relationships have not been established. Published literature shows many combinations in respect of oil, polyol and fatty acid type. Alkyds with long oil lengths often show greater initial flexibility but are more prone than shorter oil alkyds to rapid change on weathering. There are proprietary alkyds which have been formulated specifically for exterior wood. Major manufacturers also provide emulsified versions which may be used alone or blended with acrylic resins.

Urethane alkyds have specific advantages in terms of quicker surface and through dry, but this is accompanied by a decrease in extensibility which invariably leads to a greater incidence of cracking and flaking on exterior wood. The same problems beset silicone alkyds which show

Table 7 Summary of differences between solvent- and water-borne gloss paints

	Solvent-borne gloss paints	Water-borne gloss paints
Advantages	High initial gloss and distinctness of image, bright colours	Low VOC
	Good flow and levelling	Not flammable
	Long open time	Quick dry and re-coat in good conditions
	Easy repair and sanding	Low odour
	Dry under adverse weather conditions, including high humidity	Easy clean up
	High build (from high solids)	Tough flexible films with good exterior durability
	Deep penetration	Non-yellowing
		Good gloss retention
Disadvantages	High VOC	Poor drying under adverse conditions
	Flammable	Short 'open time' (wet-edge lost)
	Slow to touch dry stage	Lower build (from lower solids)
	Strong initial and post-odour	Lower initial gloss
	Yellowing tendency	Lower optical efficiency (opacity)
	Poor exterior gloss retention	Relatively poor flow
	Solvent clean up	Difficult to sand and repair
	Embrittle on exterior exposure	Prone to micro-biological attack
		Dirt pick-up and blocking
		Poor penetration into wood

Note. The perception of these differences depends on individual users and groups (e.g. trade vs retail) and also varies between countries.

excellent gloss retention but reduced extensibility. Silicone alkyds have outstandingly good chalk resistance but although this may seem an advantage it can result in very high dirt pick-up as there is no self-cleansing action. This highlights the careful balancing of properties that must be made in developing an exterior coating. The choice of TiO_2 grade can have a significant effect on chalking and considerable data have been published by major TiO_2 manufacturers.

Lower gloss finishes for exterior wood are usually formulated to be applied direct to bare wood or over an appropriate primer. To some extent, they combine the function of both undercoat and finish. They do not form a very clearly defined product group and are sometimes described as 'opaque stains' or by proprietary names. Ideally, they should be based on a fairly permeable binder and, like gloss finishes, be flexible and fungicidally protected. Choice of matting agent is critical as it is necessary to raise the PVC but without sacrificing mechanical properties. Pigments with a high oil absorption will, lower the CPVC, in effect reducing the availability of 'free' binder. To partly compensate for any loss of extensibility in this way, extenders with a reinforcing effect such as talc or mica may be employed.

Water-borne acrylic gloss finishes suitable for exterior woodwork are well established in many parts of the world, especially where large areas of wooden cladding are found. A major advantage, compared with alkyd-based paints (Table 7), is the good extensibility which is maintained for long periods of weathering. In one notable case, a maintenance free period of up to 28 years was reported [3]. In this case, particular attention was paid to achieving a high film thickness and the subsequent failure mode was very slow erosion rather than flaking.

Advances in the development of new thickeners, such as the associative types, have greatly improved the flow and levelling properties of water-borne finishes, though considerable optimisation is required to obtain a balanced formulation for a specific group of raw materials. Although the highest initial gloss levels and distinctness of image achieved by alkyd systems is not fully matched by water-borne systems, the latter normally show a much slower rate of gloss loss, and are generally superior after a period of weathering. However, the thermoplastic nature of acrylic dispersion polymers means that dirt pick-up is higher. Dirt becomes ingrained into the film itself and becomes very difficult to remove. This problem was more apparent when gloss finishes were based on the same dispersions designed for semi-gloss finishes where the higher level of pigment effectively reduces thermoplasticity. More recently, latexes based on harder but still flexible resins have become available base on core–shell and other morphologies. This effectively reduces the problem. A related problem is that of 'blocking', which can be caused by a soft polymer and will be exacerbated by water-soluble material being re-solubilised. A major problem area is the space between rebate and opening window lights. This problem is also reduced by the newer generation latexes available, but care in the selection of other ingredients must be exercised. Blocking resistance is sometimes improved by the incorporation of a proportion of non-film-forming hard latex which, in formulation terms, is considered as part of the extender PVC.

Another characteristic problem with dispersion gloss finishes is a short open time, especially in windy conditions; clearly the effect of this on appearance will be more noticeable with a gloss, as opposed to mid-sheen finish. As a consequence, water-borne latex coatings will often contain high amounts of co-solvent. It is factors such as this, combined with practical considerations, such as application in damp or cool weather, that have held back the market penetration of acrylic gloss paints. Solvent-borne alkyd gloss paints thus remained the preferred choice in many countries. However, new developments combined with more stringent solvent emissions legislation will continue to change the balance. Erosion of the alkyd gloss sector has been countered with developments in high solids alkyds, and in alkyd emulsions.

2.3.8. Alkyd developments

Continued use of alkyds in some sectors is contingent on lowering the VOC; the precise level required will depend on the product category. The two main choices are:

(1) High solids alkyds (see also Section 2.4.7 of chapter 3)
(2) Water-soluble or emulsified alkyds

Both routes have been extensively explored (e.g. [4]). The route to high solids requires a change to molecular weight and molecular structure. In principle, a reactive diluent might be used alone or in combination with another binder such as an alkyd. Alkyds typically have a molecular weight in the range 40,000–100,000 which is too high to give fluid systems even when diluted with a reactive diluent. High solids alkyds are formulated to have a much lower molecular weight in the range 12,000–20,000. Manufacture will be typically from highly branched fatty acid esters with sufficient residual hydroxyl functionality to react with a poly-functional acid. They can be used to formulate to around 200 g l^{-1}. Dendritic resins of even higher branching have also been used as an approach to lower solids.

Alkyds can also be emulsified using either internal or external stabilisers. Differences in the synthesis route are required and the overall process is usually one of phase inversion. Water acts as a chain transfer agent (affecting autoxidation) and can increase drying times. Usually, a different drier balance is required. Drying can be improved using a shorter oil-length alkyd but this may reduce extensibility. A solution is then to blend with acrylic, or other, polymeric dispersions to produce a 'hybrid' combination which may have specific advantages for wood coating by combining better penetration with a tough polymeric film (Table 8).

Lower gloss water-borne finishes (i.e. mid-sheen) are readily developed from either similar or softer polymer dispersions than those used for gloss finishes. In comparison with solvent-borne mid-sheen finishes, it is

Table 8 Summary of options for exterior wood finishes

Coating type	Notes
Solvent-borne 'high' solids alkyd	– Feasible with minor re-formulating; some colours are more difficult than others – High solids paints are inevitably more expensive than lower solids and may not seem competitive for customers – Might be seen as out of step with lower VOC options – Capable of high gloss, but thick films may dry slowly
Water-borne vinyl/acrylic or all acrylic dispersion (latex paint)	– Gloss paints based on acrylic latexes are well established and will use associative thickeners – Careful formulation is required to get best gloss but will not match gloss, flow and build of solvent-borne system above – Good durability
Water-borne alkyd emulsion	– Some benefits of solvent-borne but overall lower solids – VOC below 50 g l^{-1} – Better penetration than water-borne dispersions – May need shorter oil alkyd and therefore prone to embrittlement
Water-borne alkyd emulsion/acrylic dispersion blend	– May not achieve the highest gloss and build, but some synergy between two technologies and provide a good compromise – Care needed in tinting and order of addition in manufacture

easier to compensate for the consequences of a higher PVC on mechanical properties as the binder is not undergoing oxidation.

The higher permeability of water-borne dispersion paints is sometimes seen as an advantage for wood coatings, but for some situations it can prove too high (leading to excess movement) and pigmentation or choice of polymer should be adjusted accordingly. Surface moulds may grow more readily on water-soluble components in the coating and it is essential that they are fungicidally protected.

3. DECORATIVE COATINGS FOR INTERIOR WOOD

3.1. General purpose

The main bulk of decorative paint for interiors is for walls and ceilings, which are predominantly, plaster-based substrates. Interior timber is found in areas such as doors, skirtings (profiles or 'trim'), railings and banisters. These have been traditionally painted with both opaque and transparent systems at a variety of gloss levels, including a full gloss. Opaque paints have been largely based on solvent-borne alkyds; the formulations were of a general purpose nature. In contrast to exterior products, more emphasis is placed on appearance and quick drying; products would be less flexible than their exterior counterparts. A problem with all solvent-borne alkyds used indoors is the high primary odour from the solvent and the secondary odours caused by autoxidation. Thus even before the present legislation there was a move towards water-borne products which was aided by developments in latex and associative thickener technology. The 2010 phase of the VOC in Paints Directive (2004/42/CE) will put further pressure on the transition from solvent to water-borne for interiors, though it should be noted that the limits for interior and exterior trim are less onerous that for broad wall finishes, the use of high solids ($<300 \text{ g l}^{-1}$) solvent products is therefore not ruled out.

The situation for interior varnishes is similar. Formulating a glossy water-borne varnish that will match the appearance of the higher solids solvent-borne product remains difficult. However, polyurethane dispersions have gone a long way to close the gap and blended with acrylic dispersions can be formulated to give matt, silk and gloss products. They are quick drying and can be formulated to give hard films.

3.2. Wood flooring

The flooring market covers many substrates, products and technologies. Carpets and ceramic tiles form major sectors, but there is also a significant area of wood flooring. Solid wood flooring goes back centuries and has always been popular in some countries. In the UK there has been a revival of interest in wood flooring including parquet and laminate.

Coating of wood flooring in situ covers domestic DIY renovation, but also includes professional trade products which are more specialised. The latter includes flooring for school halls, sports halls but does carry though to domestic flooring. Specialist applicators also provide minimal-dust sanding and preparation services.

The prime requirements for wood floor coatings are:

• Resistance to high levels of abrasion
• Chemical resistance (cleaning materials and spillages)
• Black heel mar resistance

Solvent-borne OMU (oil-modified urethanes) have been widely used but VOC legislation is shifting the balance to water-borne. There is increasing interest in two-pack water-borne polyurethane emulsions but advances in polyurethane dispersion technology are also being exploited. Factory coating of flooring is described in the next chapter.

4. DURABILITY OF EXTERIOR WOOD COATINGS

Like other coating types, wood coatings are judged by their perfor-mance and appearance characteristics in relation to the period over which these attributes last. Thus durability, or 'service-life performance' are important value criteria; it is axiomatic that users, and therefore producers, will wish to maximise the lifetime of coating systems. In this context, 'lifetime' refers to the period before maintenance becomes neces-sary. There is a secondary proviso that maintenance should be as easy as possible without the need for expensive removal of coatings that have partially failed.

Previous chapters have outlined the special characteristics of wood, and the factors that will inevitably cause deterioration. Properties that resist these factors such as UV resistance, extensibility and good wet adhesion have been indicated, and in turn these must be linked to the formulation and the nature of the binder. Despite these indicators, it is impossible to say *a priori* whether a specific wood coating system will have a long service life [5]. Determining or predicting service life requires testing which may include both property measurement and exposure testing. The performance specification for exterior wood products EN 927-2 requires meeting specified criteria in a Natural Weathering Test (EN 927-3). However, the latter test is an exposure period of only 1 year on a single wood species and a laboratory prepared panel; it can only repre-sent a minimum standard in relation to the many types of wood and building configuration that are experienced in practice.

It is important to stress that in isolation coatings, however good, cannot ensure maximum life for timber components. This can only be achieved by an alliance between good design, appropriate preservative treatment and the most suitable protective coating. Failure to address all these issues can lead to failure and expensive litigation [6].

4.1. Design factors

Good design includes selection of the appropriate timber, attention to detailing and the adoption of good site practices. It is outside the scope of this book to consider design aspects in detail but a few pointers to some of the more important aspects are appropriate.

4.1.1. Joinery

Design considerations with serious implications, from the point of view of decay, include the nature of joints, profiles and the way the wood is handled and fixed on site. Water is most likely to enter at joints and exposed end grain. Moisture uptake through end grain (i.e. through transverse sections) is very much greater than other faces. Work carried out at the Building Research Establishment [7] and elsewhere has shown the moisture content of un-coated wood with sealed end grain is, on average, lower than that of coated wood with unsealed end grain. This arises because joint movement can break the protective coating seal, allowing access of water which cannot readily escape. It underlines the need for joints and any exposed end grain to be sealed using boil-resistant glues and an impermeable end grain sealer.

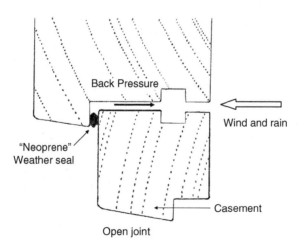

Figure 4   Window frame profile.

Two-pack polyurethane fillers are particularly effective but a number of other materials may be suitable [7].

Flat surfaces allow water to collect against joints and window profiles and must be designed with a run-off angle of 10–20°. Sharp corners are a frequent source of coating failure and should be rounded. The water resistance of windows is greatly improved by weather sealing using, for example, neoprene [Fig. 4] or PVC weatherstrips. Where timber is in contact with brick or blockwork, a damp-proof course should be provided [8].

4.1.2. Cladding

Properly designed timber cladding should make provision for both move-ment [Fig. 5] and ventilation. Coming configurations include tongue and groove, or overlapping and it is important that the fixings do not prevent sliding movement. Failure to allow for movement leads to warping, while inadequate ventilation can lead to dangerously high moisture levels being reached. Trapped water will support decay and has a generally disruptive effect on film-forming coatings. Cladding is vulnerable to moisture from outside and inside a building. A breather paper should be positioned immediately behind the cladding and a vapour barrier behind the internal lining.

4.1.3. Glazing

Paint and other coating failures are often localised at the joint between timber and glass, underlining the care that must be taken to ensure coating and glazing methods are compatible. A wide variety of glazing combinations is available, but for wood windows the glass normally sits in a rebate, with or without a bead. A long established method of glazing low-rise buildings which are to be painted, is to use linseed oil putty which when properly painted, can give adequate service, but putty glaz-ing is not suitable for frames that are to be stained, varnished or coated with water-borne paints. The appearance will be marred and the putty will not be protected, leading to early failure. Bead glazing is generally recommended instead, but where sealants are used their compatibility with coating should be confirmed. Modern windows will usually be double glazed with the lights fixed in preformed gaskets made from neoprene and other durable materials.

4.2. Preservation of timber

Timber species vary in their resistance to decay, with nearly all sapwood vulnerable. Heartwood of some species, such as oak and teak, are very resistant but ash and beech have little resistance (Table 9).

A widely accepted classification divides timber into five grades in increments of 5 years.

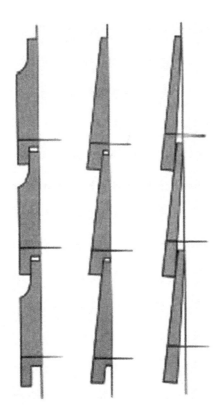

Figure 5   Three different cladding profiles.

Table 9 Durability of heartwood

Grade of durability	Approximate life in ground contact (years)	Examples
Very durable	>25	Teak, Iroko
Durable	15–25	Utile, Oak, Yew
Moderately durable	10–15	African Walnut, Douglas Fir
Non-durable	5–10	Scots Pine, Elm, Spruce
Perishable	<5	Ash, Beech, Birch

Note. All sapwood is assumed not durable.

Hazard classes are defined in European Standard EN 335-1 (*Durability of wood and wood-based products – Definition of use classes – Part 1: General*). Preservation should be considered for all sapwoods and non-durable heartwoods, where the equilibrium moisture content is likely to rise above 20%. Such a situation is likely where ventilation is poor, where the timber is in ground contact and where design features allow contact with water (Table 10).

Preservation is essential where insect or fungal attack is endemic. EN 599-1 specifies for each of the hazard classes defined in EN 335-1, the minimum performance requirements for wood preservatives for the preventative treatment of solid timber against biological deterioration. Since it is the role of preservatives to be toxic, they present significant environmental problems and are the subject to new legislation in both composition and application. Within Europe, relevant legislation includes the Biocidal Products Directive and the Environmental Protection Acts.

The three main established classes of preservative are:

(1) Tar oils
(2) Water-borne preservatives
(3) Organic solvent preservatives

4.2.1. Tar oils

Tar oils are typified by creosote which may be derived from coal or wood distillation. Timber treated with creosote is not suitable for painting but can be stained, preferably after a period of weathering. Legislation has restricted the use of tar oil-based preservatives on the basis of carcinogenic potential. Directive 76/769/EEC and its amendments prohibit the marketing of wood treated with creosote and certain other substances. However, the directive also provides for certain specific exemptions.

Table 10 Biological hazard classes from EN 355-2 (durability of wood and wood-based products – Definition of use classes – Part 2: Application to solid wood)

Hazard class	Situation	Exposure	Wood moisture content
1	Above ground (covered – dry)	Permanently dry	Permeability < 18%
2	Above ground (covered – wetting risk)	Occasionally wet	Occasionally > 20%
3	Above ground (not covered)	Frequently wet	Frequently > 20%
4	In contact with ground or fresh water	Permanently wet	Permanently > 20%
5	In salt water	Permanently wet	Permanently > 20%

4.2.2. Organic solvent preservatives

Organic solvent-carried preservatives include pentachlorophenol, tributyl tin oxide and copper and zinc naphthenates as active ingredients, carried in hydrocarbon. The two former are under increasing environmental pressure, and as a group they are flammable with a high VOC. Additives include waxes, oil and resins. The labelling of wood preservatives in Europe is described in EN 599-2. Because hydrocarbons do not interact strongly with wood they penetrate deeply, and such preservative established a dominant role for industrial pre-treatments in the UK and elsewhere. However, it should be noted that preservative practices differ widely around the world influenced by climatic differences, different building practices and the availability of more durable wood species. It could be argued that past reliance on high standards of preservation has led to some neglect of design aspects leaving wooden joinery vulnerable to displacement by alternatives such as UPVC.

4.2.3. Water-carried preservatives

Water-soluble preservatives have long been available based on copper, chrome and arsenic compounds, with sodium dichromate as a fixative. Ammoniacal/amine copper systems provide an alternative to chromium for fixation. With quaternary ammonium compounds, fixation may be through the acidic and phenolic groups of lignin. An alternative waterborne type employs disodium octaborate which is not fixed and must be protected during and after installation. Fluoride salts may be employed for in situ remedial treatment. More recently and in response to the environmental pressures, there has been a renewed interest in water-carried preservatives. The challenge has been to overcome the poorer penetration of active ingredients, and the dimensional changes caused by water. This has been partly met by emulsified systems. Coarse macro-emulsions are generally less effective than solvent-borne equivalents but show shown some efficacy against beetle emergence. Finer micro-emulsions are more effective, and can show similar penetration depth to solvent-borne albeit at lower active ingredient concentration. Although used more for remedial purposes, they are being developed as replacements for solvent-based products in industrial applications [9].

Preservatives are applied by a variety of methods which include brushing, spraying, dipping and more effective, double vacuum or vacuum pressure processes.

Although most preservative manufacturers attend to compatibility problems between treated wood and paints, glues or glazing compounds, it would clearly be unwise to assume compatibility in every case. Formulating chemists should be alert to possible problems of inter-coat adhesion especially of water-borne coatings over water-repellent preservatives.

Other compounds, including some of the copper-based ones, can inhibit auto-oxidation.

In the UK, specification of preservatives was focused on the treatment process and the type of preservative in relation to the durability class of the timber and service environment. This has been displaced by new European Standards, in response to the Construction Products Directive. The European Standards are performance rather than process based and relate to defined hazard classes. Relevant standards are EN 350-2, EN 335-1 and EN 335-2.

4.3. Specification of exterior wood coatings

The term 'specification' is used to tightly define a product or process to an agreed standard of quality. Specification will therefore be part of a contract, and the degree of conformity may be an issue in any subsequent dispute procedure. In the main specifications apply to the trade sector of the architectural market, particularly on larger contracts. Specifications will often cross reference to National and European Standards. The specification may refer to the properties of a product, for example, in terms of performance, SHE properties, etc., or the total process for preparing and applying specified products.

Each country differs in the type of specification required. In some cases, there are legislative requirements, while in others it will depend on the detail and procedures required by the user. In the UK building industry, a widely used system is administered by the NBS (National Building Specification; http://www.thenbs.com/corporate/about.asp), there are versions suited to specific industries. NBS has been taken up by a number of coatings manufacturers and adapted to their particular product range. NBS provides a framework within which a specification can be drafted it is supported by a library of clauses and guidance notes. Within the overall framework, there is flexibility to adapt to specific needs in areas which include the following.

4.3.1. Choosing the correct types of coating system
The content of specification may differ according to whether it is drafted by the user or supplier, and according to the responsibilities of the applicator and manufacturer. In very general terms, a specification must be clear about how the correct coating system is to be chosen. For a single manufacturer with a range of alternative products, there will be a set of phrases describing different system combinations. Where there is a choice of products (e.g. between water- and solvent-borne, or system combinations), then clear guidelines on product selection must be given. This is a decision that might be taken by the user or specifier. In some cases, alternative specifications (and costings) may need to be supplied.

4.3.2. Condition of surface

This is clearly a major factor which will determine the preferred system for the chosen technology. In factory-applied coatings (industrial joinery), it is probable that the substrate will be new wood. This is less likely on site where maintenance painting is more common. However, notwithstanding the advantages of factory-finished joinery, there will be occasions where wood is in the primed or finished state but with little or no exposure.

Under this heading, it is necessary to specify, for example:

- Interior or exterior
- New or un-coated
- Previously decorated (maintenance or other change of existing work)
- Primed or sealed
- Preservative treatments

Supplementary information will be necessary, for example, to allow for the nature of any preservative treatment. For example, water-repellent treatments are compatible with water-borne coatings.

4.3.3. Nature of substrate

The specification must consider suitability for different wood substrates and derived products, for example:

- Non-resinous softwood
- Resinous softwood
- Hardwoods
- Specific woods with known problems, for example, oak, western red cedar
- Sawn wood
- Composite boards (e.g. MDF, plywood)

BS EN 350-2: Part 2 gives some guidance on timber species.

4.3.4. Surface preparation

Different techniques will be required to prepare surface for maintenance (see also maintenance section); methods will cover:

- Bare un-weathered timber
- Factory stained or primed timber
- Painted timber to be over-coated with wood stain
- Removal of old coatings

In some tendering procedures, a sample to the quality required may be provided and the specification must be drafted in appropriate terms.

4.3.5. End-use categories
The need to control moisture movement, as discussed in Section 2, is an important factor in selection and specification:

- Stable – for example, window joinery
- Semi-stable – for example, cladding
- Non-stable – for example, fencing

4.3.6. Severity of exposure
The degree of exposure to the climate will strongly influence the choice of system, and the number of coats that are to be applied. Frequency of maintenance may also be noted here.

4.3.7. Site and work instructions
Control of spreading rates is essential to coating system performance in order to achieve required film build. Methods of inspection may need to be specified [10].

4.3.8. Clauses about compatibility with other products
There are a number of general instructions about storage of material, safety (COSHH) that can be covered under this broad heading.

4.4. Maintenance of exterior wood coatings

Much testing and development of coating systems take place using new wood under ideal indoor conditions. This approach is ideal for coatings that are to be applied under controlled factory conditions but is less relevant for coatings that are to be applied outdoors, or used in maintenance painting. The effect of both weather and variable substrates need to be taken into account in what is a sizeable part of the exterior maintenance market. The expression 'pampered panels' has been used to describe test panels prepared under laboratory conditions, recognising that external painting can be more demanding. The use of laboratory tests is of course justified for reproducibility and repeatability reasons, but there is the possibility of performance reversals on different substrates, or operational consequences that are outside the scope of the test.

The implication of operational factors such as the weather was noted earlier. Certainly, the preference by some trade applicators for solvent-borne paints reflects the latter's greater robustness in application during adverse weather, irrespective of subsequent appearance and performance. Some maintenance specifiers have had to review a decision to use only water-borne paints (for environmental reasons) as this severely curtailed the painting 'season'.

The following is a general checklist that might be used prior to exterior painting:

- Timber to be coated is appropriate for the job, has the correct moisture level and is not too resinous.
- Timber is free from defects, such as evidence of bacterial or biological attack.
- Knots are sealed with an appropriate treatment.
- Appropriate preservative treatment has been applied and will not react adversely with the intended type of surface coating.
- Components are of suitable design with regard to profiles, drainage, etc.
- Correct procedures have been applied during fabrication of the piece, particularly for joints, end grain sealing, etc.
- Timber components have been correctly handled in transit and stored correctly on site.
- For windows, the glazing system is compatible with the chosen coating system.
- Application conditions (temperature, precipitation) are within the recommended range.
- A correct distinction is drawn between bare wood, factory-primed work and redecoration.

4.4.1. Redecoration and maintenance
Operations involved in maintenance include:

- Removal
- Cleaning
- Sanding and rubbing down
- Knotting
- Stopping and filling
- Bringing forward
- Joint repair

4.4.2. Removing coatings
Coatings that are in sound condition (free from cracking and excess chalking, firmly adhering) do not need removal. Minor damage can be repaired. Exceptions to this would include a major change of appearance (e.g. opaque to transparent) or technology. Where coatings are to be removed, there is evidence [11] that the method of removal can affect subsequent durability. Methods of removing coatings include:

- Chemical removal
- Burning off
- Heat treatment with mechanical scraping
- Mechanical sanding
- Sanding

4.4.3. Chemical treatments

Paint strippers based on methylene chloride are messy to use on in situ coatings; the same applies to strong alkali removers such as caustic soda which is widely used for cleaning up old furniture. Chemical treatments are also suspected for weakening the interface and in the case of strong alkalis can discolour some woods.

4.4.4. Burning off

The familiar propane-fuelled blow torch is widely used to remove heavy build up traditional oil-based coatings, but even in the hands of a skilled operative it will cause some charring and is less suited if transparent coatings are to be applied. Burning is more effective on materials that char and much more difficult to use on thermoplastic paints. Burning methods do not give a truly clean surface; there will be residual paint in pores and crevices. Blow torches present a serious fire hazard if carelessly used; denatured wood will sometimes smoulder beneath the visible surface.

4.4.5. Hot-air strippers

These are a milder version the above; the lower temperature reduces the risk of scorching but they are slower and unable to cope with all types of coating. They are more suited to DIY than professional use.

4.4.6. Mechanical sanding

Orbital or belt sanders are a fast method of removing the bulk of a coating though clearly creating a lot of dust. Masks should be worn and if lead-based paints are suspected, then dust collection is required.

4.4.7. Sandblasting

This is the fastest method but is more capital intensive and probably only worth consideration for larger jobs. Sandblasting is the most effective in removing all traces of a previous coating leaving a good base for stains or opaque coatings. It is advisable when using sandblasting to first check on a small area that the fabric of the building will not be damaged.

It has been observed that cleaned up surfaces do not usually perform as well as new wood [11]. Paint systems with an intermediate level of durability were particularly affected with sandblasting giving the best results with regard to subsequent flaking failure, followed by mechanical sanding and then burn-off. Cracking was a problem with all the clean-up methods including sandblasting. It might be tentatively assumed that flaking was influenced by adhesion, with sandblasting giving a cleaner surface, whereas cracking was more influenced by the permeability of the coating.

4.4.8. Cleaning

Where coatings do not need to be removed, then cleaning is still required to remove dirt, oil, grease, wax or other contaminants. Washing with detergent of soap solution may be combined with light abrasion. Solvent cleaning may be necessary where surfaces are contaminated with oil or polish. A sterilising wash is useful if there has been biological contamination, diluted bleach is cost effective, appropriate eye and skin protection should be used.

4.4.9. Sanding and rubbing down

This can be part of the cleaning process and is used to improve appearance where a high gloss is desired. Inter-coat adhesion can be applied by sanding and this is certainly the case if the wood surface is denatured.

4.4.10. Knotting and tannin staining

Resinous exudations from knots are a perennial problem with some wood species and will discolour paints. The traditional treatment is two coats of shellac; aluminium primers can also be effective. Tannin staining can be a major problem with water-borne paints (see discussion under Section 2.3.5).

4.4.11. Stopping and filling

By tradition, nail and screw holes, open joints, deep cracks and similar imperfections are 'stopped' using a fairly stiff material, while shallow depressions, rough or open grain, fine crack, etc., are 'filled' using a more fluid material. Deep holes should be filled in layers to avoid excess shrinkage; major types of stoppers and fillers are:

- Oil-based
- Methyl cellulose/gypsum (powder form to be mixed with water)
- Emulsion-based (usually ready for use)
- Two-pack fillers (often peroxide-activated polyester, or epoxy amine systems)

4.4.12. Bringing forward

This term is used in repair, old paint is 'feather edged' and bare area built up to the original film thickness.

4.4.13. Joint repair

Open joints should not be filled without first opening the joint to a 'vee'-shaped groove. Any exposed end grain should be treated with a preservative primer and sealed.

4.4.14. Operational costs of maintenance – Coating type

Taking into account differences in the cost of preparation, application and materials, the initial cost per unit area of painting (or varnishing) is likely to be about twice that of wood stain treatment. But when maintenance is taken into consideration, wood stains usually need treating more frequently; typically at intervals of 2–3 years, compared to four or five for paints. But this is offset, at least in part, by the fact that less preparation and fewer coats are usually needed to maintain wood stains than to maintain paints. The advantage probably lies with stains for cladding and paint for window joinery.

European Standards Cited

- BS EN 927-1:1997; Paints and varnishes. Coating materials and coating systems for exterior wood. Classification and selection
- BS EN 350-1:1994; Durability of wood and wood-based products. Natural durability of solid wood. Guide to the principles of testing and classification of natural durability of wood
- BS EN 350-1:1994; Durability of wood and wood-based products. Natural durability of solid wood. Guide to the principles of testing and classification of natural durability of wood
- BS EN 599-1:1997; Durability of wood and wood-based products. Performance of preservatives as determined by biological tests. Specification according to hazard class
- BS EN 335-1:2006; Durability of wood and wood-based products. Definitions of use classes. General
- BS EN 335-2:2006; Durability of wood and wood-based products. Definition of use classes. Application to solid wood
- BS EN 335-3:1996; Hazard classes of wood and wood-based products against biological attack. Application to wood-based panels
- BS EN 15434:2006; Glass in building. Product standard for structural and/or ultra-violet-resistant sealant (for use with structural sealant glazing and/or insulating glass units with exposed seals)

REFERENCES

[1] Schaller, C., and Rogez, D. (2007). New approaches in wood coating stabilisation. *J. Coatings Technol. Res.* **4**(4), 401–409.
[2] Twene, D., and Mestach, D. (2002). Novel approach in preventing the migration of tannins in water-borne wood applications. *In* "Proceedings of the PRA 3rd International Wood Coatings Congress 'Wood Coatings: Foundations for the Future'", The Hague, 17 pp (paper 44).

[3] Miller, E. R., and Jones, T. (2000). Maintenance-free coating of wooden windows. *In* "Proceedings of the PRA 'Wood Coatings Challenges and Solutions in the 21st Century' Conference", The Hague, 16 pp (paper 7).

[4] Sorensen, K. (2008). Routes to volatile organic compound (VOC) compliant alkyd coatings. *In* "Proceedings of the PRA 2nd International Waterborne & High Solids Coatings Conference 'Reducing Environmental Impact'", Brussels, 12 pp (paper 9).

[5] De Meijer, M. (2002). Mechanisms of failure in exterior wood coatings. *In* "Proceedings of the PRA 3rd International Wood Coatings Congress 'Wood Coatings: Foundations for the Future'", The Hague (paper 40).

[6] US Court of Appeal (For the Eighth Circuit). No 02-2833 & 02-2869. Filed March 23, 2005.

[7] Miller, E. R., Boxall, J., and Carey, J. K. (1987). External joinery: End grain sealers and moisture control. Building Research Establishment (UK) Information Paper 20.

[8] Baird, J. (1994). Good window detailing as a means of maximising the service life of paint finishes. *In* "Proceedings of the PRA 'Modern Wood Finishes: Meeting the Technical Challenge' Symposium", Harrogate (paper 1).

[9] Worringham, J. H. M. (1996). Water-borne preservative and allied low build coatings for wood. *In* "OCCA Symposium 'Environmentally Friendly Wood Preservatives and Coatings'", 18 April 1996 (paper 3).

[10] Suttie, E., Thorpe, W., and Dearling, T. (2005). Improving maintenance painting of exterior wood. *Surface Coatings Int. Coatings J.* **88**(A7), 277–281.

[11] Boxall, J., and Smith, G. A. (1987). Maintaining paintwork on exterior timber. Building Research Establishment Information Paper IP 16/87.

CHAPTER **8**

Market Needs and End Uses (2) – Industrial Wood Coatings

1. INTRODUCTION

In the previous chapter, the principal differences between decorative and industrial wood coatings were outlined. It was noted that many differences are operational in nature and that this leads to alternative technology choices. A further difference arises from the fact that a major part of the decorative market concerns maintenance painting. In this present chapter, industrial coatings are assumed to be applied under factory conditions;

Wood Coatings: Theory and Practice
DOI: 10.1016/B978-0-444-52840-7.00008-4

subsequently (e.g. for joinery), there may be a need for maintenance and redecoration, but this is then likely to become an in situ decorative process using the products already described.

The broad scope of the industrial sector is indicated in Fig. 1. The term 'joinery' is used rather loosely to denote fabricated wood in buildings. It will include window frames, doors and various profiles or trim, such as skirting boards. Coatings for interior components will often be formulated differently from exterior ones. However in the case of a window frame, which presents an interior and exterior surface, it may be more convenient to continue the coating all around the profile. In this case, an exterior grade of coating should be used. Coating operations will also be modified according to the finish that is required; thus, there may be differences between opaque and transparent coatings. In the case of interior furniture, some markets prefer a solid high-build coating, while in others a more 'open-pore' effect is preferred.

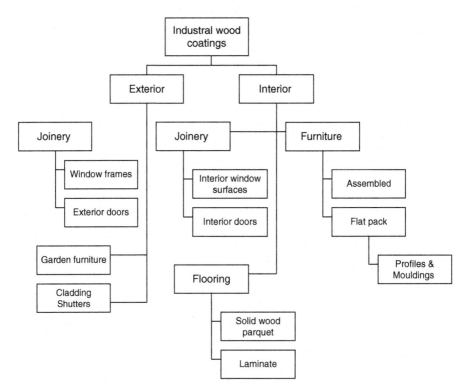

Figure 1 Scope of the industrial sector.

2. INDUSTRIAL FINISHING OF WOOD JOINERY

It is widely recognised that the best approach to joinery protection is to produce fully finished components in the factory. This can ensure that all finishing and glazing are carried out under controlled conditions to a quality-assured standard, and yields a superior quality of finish. The production of fully finished joinery is normal in several European states and is being adopted to an increasing extent by window manufacturers. This is in contrast to an earlier practice of supplying windows to construction sites in a partially finished state, usually comprising only one or two coats of primer. As noted elsewhere in this book, the performance of even the best factory-applied coating will be strongly dependent on:

- The design detailing and construction of the joinery
- Wood quality
- Wood species (including laminar wood)
- Effective sealing of joints
- The quality of the coating systems

2.1. Preservation

Although practices differ between countries, it is still common for joinery intended for factory finishing to be pre-treated with a solvent-borne preservative. It is essential that the treated components are allowed an adequate time (48 h minimum) for solvent release if adverse interactions with the coating system are to be avoided.

Water-borne preservatives are increasingly accepted. When applied by double vacuum, they introduce the risk of distortion and also a requirement for extended periods of drying. Hence, there is a trend towards the application of water-borne preservatives by flow coating. Drying times may then be as short as 3 h which facilitates flow-line production, though some risk of fibre and grain-raising remains. Such treatments are principally promoted as blue-stain inhibitors and their adequacy in decay prevention may be less. However, the need for full preservation in a well-designed window (and its surround) considerably reduces the need for preservation, since the window should not remain wet for long periods.

In the USA and elsewhere, there has been growing pressure to replace copper chrome arsenic preservatives (CCA). One contender is for 4, 5-dichloro-2-n-octyl-4-isothiazolin-3-one (DCOIT), which is the active ingredient in some proprietary preservative systems [1].

2.2. Coating systems

Although solvent-borne coating systems based on alkyds and 2K polyurethanes have been used in the past, many industrial finishing systems for joinery are now water-borne, and based on acrylic or alkyd–acrylic polymers. These resin systems have the advantages of high film flexibility and durability, but the softness of the fresh films means that finished components require careful handling. Blocking (i.e. damage caused by stacking of components) is also a potential problem. Developments in core–shell polymers have led to coatings in which the hard core imparts toughness and early block resistance, and the softer shell contributes the elasticity so important to long-term durability, and also improved hail resistance. Some woods are troublesome from the point of tannin staining in the case of Iroko extractives sometimes appear as white spots.

Improved film properties can be achieved by isocyanate-cured systems but in water there are additional problems to be overcome. The two components must combine hydroxyl functionality and appropriate isocyanate functionality in a situation where water itself will also compete for isocyanate groups producing carbamic acid, which decomposes to release CO_2. The latter reaction may be reduced by blocking but a surplus of isocyanate (NCO/OH ratio of 1.5) is likely to be required which adds to the cost.

In principle, it is possible to emulsify conventional isocyanates with the polyol component, but high shear mixing is required. It is thus preferable to use a water dispersible polyisocyanate prepared from polyethers. Both HDI and IPDI trimers have been used.

Hydroxy functional acrylic resins can be emulsified with polymeric stabilisers with a relatively low molecular weight and hence good film levelling and flow properties. Alternatively, a higher molecular weight is made via emulsion polymerisation. As these are of higher molecular weight, the cross-linked network can build more rapidly with advantages in early resistance properties. However, flow may be reduced [2].

2.2.1. Typical industrial joinery finishing schedule

- Impregnation: flow coating (or dipping) of components with water-borne preservative containing resins and fungicides:
  - Equalises differential absorption of water, ideally with minimal raising of wood fibres, though this is more of a problem than solvent-borne
  - Typical drying time 2–3 h (air-ventilated oven)
- Priming:
  - Flow coating with water-borne primer, transparent or opaque, based on alkyd–acrylic technology
  - Typical drying time 3 h
  - A tannin blocking technology should ideally be used

- Intermediate coat (optional):
  - Water-borne alkyd–acrylic coating with pore-filling properties designed to improve finish quality on coarse-textured hardwoods
- Finishing coat:
  - Airless spray application of water-borne topcoat. Resin type acrylic, polyurethane acrylic or core–shell polymer
  - Blue-stain inhibitor
  - Wet film thickness up to 300 μm
  - Typical drying time 2–3 h

Needless to say, there can be many variants of the schedule, including the possibility of some repair or retouching (Fig. 2).

It is most common for components to be made up into frames and casements prior to finishing, though some joinery companies coat piece parts prior gluing and assembly. The latter approach has a number of important operational advantages, in particular that end-grain surfaces are sealed effectively. The finishing line is more compact and lends itself to automation and other application methods. Surface preparation and, where necessary intercoat sanding of components is also simplified. The operational advantages of finishing a partly fabricated unit are also important in furniture manufacture.

2.3. Environmental legislation considerations

Installations producing joinery components are subject to the solvent emissions directive (SED) as described below in Section 3.5. Thus, pigmented coating applied by spray, curtain or dip techniques are required to have VOC content (excluding water) of below 520 g l^{-1} (see Section 3.5).

Figure 2   Assembled window frames before the coating process (courtesy: Vertek Group srl).

3. INDUSTRIAL FINISHING OF FURNITURE

Furniture manufacture represents the largest industrial wood coating sector (~60% of total). It is a complicated market using many different materials. In this chapter, the focus is on the coating of wood that is visible to the user; however, wood may also be used internally as a support for upholstered items. Furniture is also made from other materials including metal and plastic and may combine different materials. Flat panel, mouldings and profiles will be typically bases on products derived from wood, especially MDF. In the case of solid wood, manufacture will start with large sections that must be cut, shaped, jointed and finished by wood-working processes. A combination of different processes will be used since legs, tops, etc., may be made from different materials.

3.1. Coatings for wood furniture – Some functional types

Styles of furniture differ greatly around the world. There are noticeable differences between the USA and Europe and within Europe if one compares, for example, Italian and Scandinavian designs. This will influence the product types and the technologies that are used. Some combination of the following product types will form the basis of a coating process according to the nature of the substrate and the desired appearance.

3.1.1. Bleaches
Hydrogen peroxide activated with sodium or potassium hydroxide (or ammonia) has been widely used. Often, this was carried in methyl alcohol to reduce the grain-raising caused by water (grain-raising is a general problem with all water-based products). Methanol must be used with care due to its toxicity.

3.1.2. Sizes and washcoats
A thin coat of glue size was often used to raise and stiffen grain to aid sanding and also to even stain absorption on certain species of wood. Nowadays, a polyvinyl alcohol solution is more likely to be used. A similar function is performed by washcoats which are essentially a very low volume solids lacquer.

3.1.3. Stains
Stains are widely used to darken wood and emphasis grain. Shading or 'padding' stains are used to correct any lack of uniformity of colour. Other names are used include 'wiping stains' and 'vanish stains'. Stains can be aqueous or non-aqueous. The latter may also be called

'naphtha' stains, but are also based on alcohol, turpentine and mineral spirits. Unlike exterior wood stains, dyes rather than pigments are generally used in furniture stains. The non-aqueous stains are sometimes described as 'NGR' (non-grain-raising) and are usually preferred for hardwoods. Water-based stains cause greater grain-raising but sometimes lower penetration is an advantage where uniformity of colour is required. Non-aqueous stains can give patchiness under some circumstances (due to differential absorption), and in this case may be lightly pigmented.

3.1.4. Fillers

Fillers are used both to colour wood pores and to fill them, particularly where a very smooth high-build finish is required (this is not the case in all countries). Natural earth pigments are used (in the coloured variety) with a variety of binders which include oils, nitro-cellulose and alkyds. Pigments and extenders (china clay, lithopone) must be finely dispersed to penetrate pores. A wiping or 'padding' process removes excess material from between the pores. On flat stock reverse roller coating is used. Adequate drying of fillers is essential if subsequent shrinkage and other defects are to be avoided.

3.1.5. Sealers

Sealers, as the name implies, correct differential surface absorption and provide a basis for uniform top coating. They may be sanded further and should be non-blocking. Sanding sealers were traditionally based on nitro-cellulose modified with various resins. Zinc stearate may be added to aid sanding.

3.1.6. Topcoats

Often described as a 'lacquer', the traditional transparent topcoat was based on nitro-cellulose which gives an outstanding gloss and clarity of grain. Nitro-cellulose lacquers have the advantage that they can be polished. Traditional 'French polish' is based on a solution of shellac in methylated spirit. Nowadays for legislative, operational and economic reasons, many other technologies are used for both the topcoat and the other product types described above. Transparent coloured coats may be used for special effects; these may contain dyes but are more likely to contain finely dispersed pigments including transparent iron oxides. Transparent coatings will also contain flatting and matting aids when a gloss finish is not required.

Opaque finishing topcoats will be pigmented with titanium dioxide and other pigments appropriate to the colour. Gloss and sheen are controlled with extenders and other matting aids.

3.2. Influence of substrate on coating type

For opaque coating systems, the direct coating of wood-based panels is possible. Medium density fibre boards are particularly suited for moulded shapes.

In the case of transparent or semi-transparent coating systems, the appearance of the substrate has a greater influence. In these cases, the use of paper (or PVC)-faced panels is an option, especially for low-cost furniture. The use of solid wood or veneered panels also allows the attainment of an open-pore effect.

For raw particleboards, the surfaces of which have an inhomogeneous structure, the application of one or more filling coats is necessary. For moulded MDF boards, the application of an insulating primer is also suggested. Provided the primer has a low viscosity, such products penetrates the substrate, especially in the moulded areas, thus reinforcing the fibre cohesion. This preliminary treatment reduces the possible formation of cracks due to the internal stress induced by the cross-linked paint or derived from the humidity exchanges of the panel with environment.

In the case of substrates constituted of impregnated papers applied on particleboards or MDF, the initial cleaning of the surface (e.g. with acetone, but subject to VOC constraints), and the use of an adhesion promoter (primer) is always opportune. The function of this product is to penetrate in the paper pores promoting the adhesion of the following coats.

Also for porous veneered panels, the use of a primer coating, deeply penetrating into the cavities, is necessary to avoid a whitening defect in the pore caused by light scattering if the pore is not filled.

When paper or veneers are applied with thermoplastic adhesives, careful attention must be paid to the coating system used because of the sensitivity of such adhesives towards organic solvents and heat. The combination of these two factors could cause detachment of the veneer.

3.3. Coatings for wood furniture – Factors influencing technology choice

It should be emphasised that there is no single 'correct' solution for industrial wood coating. The choice of technology will depend on many factors and furniture manufacturers may reach different conclusions according to circumstances. A formal value analysis can be very helpful when considering the choice of technology from a particular perspective.

Some of the major considerations that will influence the choice of technology for furniture and related coating operations are the following.

3.3.1. Economic factors and economies of scale

Economics is always important in manufacturing. Although the price of the coating itself is important, it may be relatively minor compared to the cost of the item itself. What can be very important is the total operational cost. There will be very big differences between a small business using predominantly hand finishing techniques, and a large factory with a high degree of automation and capital intensive equipment. Some coating technologies and processes are inherently expensive and it is this, rather than the cost of the material itself which must be considered. This is also a factor to be considered when technology is changed; does it require more capital investment? The use of either UV or powder coating technologies has major operational implications.

3.3.2. Operational factors – Fully assembled or knock-down?

Important operational factors include issues such as sanding, and the method of application. What is important from the point of view of subsequent coating is whether the furniture is 'fully assembled' or in 'knock-down' form when finished. This will be dictated by a number of factors including the size of the operation and the design involved. Furniture made from flat stock is often supplied in knock-down form for assembly by the final user.

Fully assembled furniture, such as a chair, can only easily be coated by a spraying operation. Furthermore, sanding is more difficult requiring a hand operation, or abrasive brushes. This precludes some technologies which have a harder finish. In contrast, the coating process possible for flat stock includes roller coating, vacuum coating, and curtain coating (see Chapter 9). It is also much easier to carry out sanding operations, using, for example, a belt sander, on flat items; thus, harder finishes can be used.

3.3.3. Appearance aspects – Open or closed pore?

General factors influencing the appearance of coatings were described in Chapter 6 and include colour and gloss; these will affect the formulation detail irrespective of the binder technology. Another important aspect of appearance particularly for clear and semi-transparent finishes is the desire for a 'closed-' or 'open-pore' effect.

Some woods including mahogany, walnut and oak have very obvious pores which remain visible on the surface as an indentation after coating. This is called an open-pore effect. It can be seen with many types of wood to different degrees. Coating systems can be designed to fill the pores (sometimes called a 'fully choked' effect), using fillers and high-build finishes. Alternatively with lower solids systems, the pores can be left as a feature giving a softer appearance, which is also modified by the chosen gloss level. National preference for appearance in this respect does vary. As a generalisation, it can be said that northern Europe prefers the open-pore effect, while in southern Europe a closed-pore higher build is preferred. Intermediate effects, known as 'semi-open pore', may also be specified. The consequences on the choice of technology are significant. One solution to meeting VOC legislation for established solvent-borne technologies is to raise the solids content. However, this makes it more difficult to control film thickness during application, and hence more difficult to achieve an open-pore effect. This is also true for inherently high-build technologies such as radiation curing. Consequently, it may be necessary to dilute the oligomeric binders with solvent or preferably with water. The latter option is of growing interest but imposes an additional drying stage on the coating process before the UV cure.

3.4. Functional needs

In addition to the desired appearance, no choice of technology for furniture can be made without considering the functional requirements. Market sub-sectors include, for example:

- Home
- Kitchen
- Seats
- Office
- School
- Contract

Each will have some different requirement. For example in kitchen cabinets, the finishes will have to resist a wide variety of substances including hot fat and food stains. Tables in domestic situations must also resist alcohol (from drinks) and the many cleaning sprays that are used to wipe and disinfect surfaces. The requirements are more stringent in the catering sector. In general flat surfaces, for example, tables are more challenged than vertical surfaces and the finishing system may be adjusted accordingly. Within the broad areas of technology described elsewhere such as polyurethane and polyester, it is possible to make many variations according to detailed chemistry. The selection of technology for a particular application will often depend on performance against a selection of test methods appropriate to the intended application (see Chapter 6).

3.5. Legislation

All the coatings described in this chapter are subject to various pollution control acts and the mandatory requirements of the EU SED. Implementation of regulations, and the capacity of the unit to which they apply, may differ in some respects between countries and for convenience reference is made here to Process Guidance Note 6/33(04) issued by the UK Department for Environment, Food and Rural Affairs (Defra) which covers wood coating processes and installations. It is based on an EU consensus on best available practice (BAT). It should be noted that the detailed implementation of the legislation is complex and beyond the scope of this chapter. There are constraints on both VOC and non-VOC emissions (e.g. carbon monoxide, isocyanates, particulates). For VOC releases, there are two compliance options:

- Reduction scheme/compliant coatings
- Emission and fugitive limits:
  - *Note.* Fugitive refers to solvent that may be lost through cleaning, or disposal of used containers, as opposed to the emission from the work piece.

Compliance with emission limits is aided by various abatement techniques and requires a solvent management plan; however, wood coating installations do not generally use this option. This is in contrast with coil coating where abatement is the main strategy. The published VOC limits (Table 10 in PGN 6/33) do not apply when compliant coatings are used. Compliant coatings (see Chapter 3) will include:

- Water-borne coatings (with a low VOC level)
- Higher solids coatings
- Powder coatings
- Organic solvent free liquid coatings:
  - Typically radiation cured

The maximum VOC for compliant wood coatings were specified in 2004 as given in Table 1.

Manufacturers may wish to calculate VOC emissions for other purposes or to meet local regulations. To do this would require:

- The VOC content of the coating product as a percentage (%)
- The spreading rate (application rate) in g m^{-2}
- The transfer efficiency (%)

This will enable the VOC emitted to be calculated in g m^{-2}; it must be derived for each part of a multi-coat system. Any recovery of waste, ideally with recycling, will improve the overall efficiency:

- It is recommended that the VOC concentration is measured, rather than calculated; a method is given in Section 10 of PG 6/33(04).

EXAMPLE

If a water-borne coating has a VOC content of 5% by weight and is applied at 250 g m^{-2}, then the emitted VOC is $250 \times 0.05 = 12.5$ g. But at a transfer efficiency of say 50% this would become $12.5 \text{ g} \times (100/50) = 25$ g m^{-2}.

Table 1 Maximum VOC for compliant coatings (UK 2004)

Group	Coating type	VOC (g l^{-1}) but not including water[a]
a	Fillers	370
b	Clear coating applied by vacuum or roller coating methods	220
c	Pigmented coating applied by vacuum or roller coating methods	265
d	Pigmented coating applied by spray, curtain or dip techniques (except group g)	520
e	Clear coating applied by spray, curtain or dip techniques (except group g)	475
f	Clear coating applied by spray, curtain or dip techniques where all other coats are water-borne coatings containing no more than 10% by weight of VOC	600
g	Pencil end dipping lacquers	650

[a] The VOC is based on the volume from any water deducted; this is more severe than the calculation used in the decorative sector and will give a higher VOC.

3.6. Coatings technology for furniture

Some indication of the broad distribution of furniture coating technology within Europe was given in Table 4 of Chapter 1. The actual detail is more complicated and can only be interpreted in the context of the legislative, operational and other factors noted above. Both the assembly and appearance cause operational consequences (Fig. 3).

3.6.1. Assembled furniture

3.6.1.1. High solids technologies   The long-established process is by hand spray using an easily applied coating system such as nitro-cellulose lacquers or acid-catalysed alkyd amino. Operations include staining and filling followed by 2–4 coats of basecoat and topcoat with sanding between coats. As VOC contents of traditional systems are around 75%, the total emissions will be high and a continued rapid decline of nitro-cellulose lacquers is inevitable.

The impact of legislation can be mitigated with water-based stains and a higher solids acid-curing UF or MF lacquer with VOC reduced to 60%. To overcome drying and blocking problems, it is normal to include polymers such as cellulose acetate butyrate (CAB), and this allows a practical drying regime at ambient temperatures. CAB is also

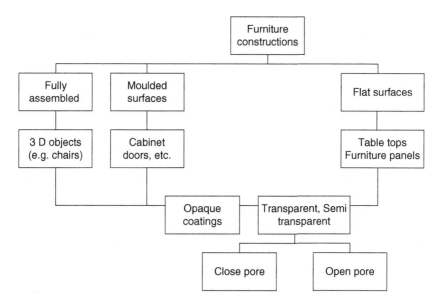

Figure 3   Furniture constructions: substrates and finishing options

used with thermoplastic acrylic resins. Polyester, polyester/melamine and polyurethane resins are also available at 60–70% solids content. It should be remarked that the situation in the USA is different for high solids systems in that the legislation enable the use of 'exempt' solvents including ketones which can be excluded from the VOC calculation.

It could be argued that high solids technology is reaching its limit for hand spraying though it has advantages for darker finishes where a glassy high gloss finish is required. At higher solids, the correspondingly higher viscosities restrict, or prevent, the option of adding modifying polymer. Blocking and dust pick-up then become major problems. Another problem with high solids finishes is that natural 'open-grain' effects become more difficult to achieve. This can be minimised with lower film weights but as the coating solids goes up control of film thickness requires constant attention if gloss variation is to be avoided.

3.6.1.2. UV technology   The problems of drying and blocking could in principle be overcome by UV curing, but such finishes tend to be hard and abrasion resistant. Abrasive brushes of the sort used to denib acid-catalysed finishes have little effect and three-dimensional objects, or profiles, become very difficult or time consuming to denib. Also, the problem of achieving open-pore effects remains, unless lower solids water-based UV coatings are used.

3.6.1.3. Water-borne coatings   Because water-borne coatings are able to combine low solids with a low VOC, they can be formulated to undergo considerable shrinkage on drying. This is beneficial if open-pore effects are required, and also to allow the effective use of matting aids where a satin or low gloss is required. However, water is unique among solvents for its affinity to biological systems and its prevalence in the atmosphere, resulting in some specific differences from non-aqueous solvents. Among these must be numbered grain-raising and drying issues.

All species of wood will absorb moisture as liquid or vapour but the effect on grain-raising is very species dependent. Oak (*Quercus robur*), for example, can show severe grain-raising whereas Pine (*Pinus sylvestris*) is much less prone. Veneers are generally less affected than the solid wood. Where grain-raising is a problem then extra coats with de-nibbing between coats is one solution. Grain-raising is also a function of wood moisture content and is less for most water-borne coatings if the moisture content is higher (10–12%) than the normal (7–8%) for solvent-borne basecoats. A light spray with water followed by drying and sanding can mitigate the effect grain-raising without the need for an extra coat. Alternatively, it is possible to combine solvent-borne basecoats with water-borne topcoats but not all VOC legislation will allow this.

Unlike their solvent-borne counterparts, water-borne lacquers will be sensitive to humidity, and a relative humidity of 50% (at 18–20 °C) will be required if drying times are not to be seriously retarded. For all but the smallest units a forced drying regime will advantageous. For the first coat on unsealed timber absorption of water will also slow the drying rate, and it might be advantageous to raise the temperature of the surface with infra red heaters; as a beneficial spin-off grain-raising may then also be reduced.

The appearance and properties of water-borne coatings will depend very much on the specific chemistry chosen. Often, the polymers used in used in water-borne dispersions will have a lower refractive index than solvent-borne polymers, and this possibly combined with a degree of incomplete coalescence gives a translucent effect especially on dark substrates. For this reason, water-borne finishes are more suited to low-build finishes on light coloured timbers. Another coloration effect arises from the fact that water-borne lacquers are almost invariably high in pH (i.e. alkaline) and can produce a significantly different colour from acid-catalysed products either initially or over a period of time. Some water-borne coatings contain amines which have specific coloration effects with certain woods.

The appearance of clear coatings on wood is sometimes described by the subjective term 'Anfeuerung' [3]. It relates to the enhancement of wood appearance in terms of grain contrast, warmth and lack of haze. Generally a 'good' Anfeuerung is harder to achieve with water-borne systems, but the effect is system specific and will depend upon the refractive index of the polymers in the coating, and also particle size in the case of dispersion systems.

In terms of dry film properties, water-borne acrylic coatings are thermoplastic and borderline for durability tests such as marking by liquids and wet heat. This can be improved upon by cross-linking variants. Water-borne polyurethane technology offers a number of techniques for achieving the necessary resistance properties.

3.6.2. Knock-down furniture

Comparison of the total VOC emissions in coating an approximately similar knock-down as opposed to ready-made piece of furniture would show the former to be considerably lower. This reflects the fact that knock-down furniture usually contains flat panels opening up the possibility of using roller coating methods of coating application which require less solvent than spray. The balance of advantages for different compliant coatings immediately changes.

3.6.2.1. UV curing   High solids UV radiation-cured coatings have become well established for flat articles including components for K-D furniture, flush doors and panels in general (Table 2). Application method is curtain coating or roller coating which has high transfer efficiency. Automatic spray is not viable without solvent addition. An advantage of UV technology is good durability and high surface hardness, though as noted above this makes sanding and de-nibbing harder and the technology cannot readily be applied to profiled furniture.

For flat stock in contrast to assembled furniture, on-line drum or belt sanders are effective. UV curing gives very rapid curing leading to fast line speeds and short drying lines. Disadvantages include the high capital investment and relative inflexibility in production; thus, the technology is most suited for the larger manufacturer with high throughput volumes. Although VOC levels are very low, there are other health and safety effects and some of the components are skin irritants.

Until recently, another disadvantage of UV curing was that pigmented products did not have practical curing times due to absorption of radiation by pigments. New lamps, catalysts and initiators have overcome this problem (at a cost) but pigmented systems are not quite as developed as their clear counterparts. It may be necessary to use low levels of photo-initiator in thick coatings; otherwise, surface absorption will block cure of the underlying portion. Polyisocyanates are sometimes used to boost curing though this will reduce pot life.

3.6.2.2. Water-borne coatings   The blocking problems mentioned above have limited the value of straight water-borne coatings for K-D furniture, particularly as high solids coatings are an economic alternative. Water-borne

Table 2 Major types of UV chemistry used in furniture coating

Type	Properties
Epoxy acrylate	Fast curing, chemical resistant, low to moderate cost
Urethane acrylate – aromatic	Tough durable and chemical resistant
Urethane acrylate – aliphatic	Non-yellowing – cost moderate to high
Polyester acrylate	Variety of types available – moderate cost
Polyether acrylic	Low viscosity and irritancy
Acrylated acrylic	Good adhesion
Novolac – acrylate	Hard, good chemical and heat resistant
Unsaturated polyester	Low cost but slow curing
Cationic epoxy	Good adhesion, no oxygen inhibition
	Sensitive to moisture and darkens after reaction
Dual cure – cationic/free radical	Good adhesion and rate of cure
	Reduced oxygen inhibition
Dual cure – free radical/ chemical	Good adhesion to porous substrates, but dark reaction

UV-curable finishes, however, offer a viable but much more costly alternative and can be applied by automatic spray, roller and vacuum coater as well as curtain coating. These coatings do not avoid grain-raising problems and cannot be exposed to the UV lamps until water removal is complete. Thereafter, the cure is rapid and energy efficient. Open-pore effects are readily achieved without the problems of bubbling that sometimes occur with acid-catalysed finishes. Film clarity is good and approaching that of high solids acid catalysed. However, it is important that no water is trapped as this can lead to turbidity where woods have deep pores.

The greater flexibility in application methods and softer finish than conventional UV cure allows profiled sections to be handled but pigmented systems are not yet fully developed. Another advantage of this technology is that overspray can be reclaimed thus raising the overall transfer efficiency and allowing some of the extra cost to be offset.

3.6.2.3. High solids coatings   Both high solids acid-catalysed alkyd amino systems, and PU, can meet current projected legislation though automatic spray is advantageous to control film weights where open-grain effects are required. After forced drying and cooling finished panels can be stacked without blocking. Thus, high solids coatings with their good interior durability are generally seen as more viable for K-D than assembled furniture and this has restricted penetration of this market by water-borne lacquers.

3.6.2.4. Powder coating   Powder coating of wood remains challenging because heating causes resinous materials to come to the surface. This is true even with UV-cured powder since some pre-heating is still required. Some success has been achieved with veneers rather than solid wood.

Powder coating of MDF has been more successful. Application is by electrostatic spray and conductivity of the surface is an important control parameter. One problem is that although heating of the work piece drives out water, and thus increases conductivity, the water is lost at the edges and a low conductivity results. A water-borne thermosetting primer coat may be used to overcome this effect; otherwise, a separate finishing operation for the edges is required.

Powders formulated for wood products are thermosetting (e.g. polyester, polyurethane, acrylic) or UV curable (polyester, acrylate, epoxy). A general problem with UV cure is that the coating will be under-cured in shadowed or concealed areas, hence the use of thermosetting powders. Powder coatings have more restricted range of gloss levels than can be achieved with liquid coatings.

ADHESION ON WOOD FURNITURE AND PANEL PRODUCTS

Good adhesion is a prerequisite for good performance in mechanical and resistance tests. Most coating processes are multi-coat and adhesive performance will be particularly associated with the first coat. Proper sanding is the first stage and with flat feed stocks aluminium oxide belts are favoured. Good adhesion is reported with both 100% UV and UV water-borne products including water-borne UV-PUD. Sometimes acrylic dispersions alone can be used. Water-dilutable resins are also a possibility including aliphatic urethane acrylates and polyether acrylates. Where water is the carrying medium, it must be largely removed before UV curing. Although some will be absorbed by the wood a pre-drying stage will be necessary. Dry air ovens may be used. Coatings must not be too thick at the drying stage otherwise bubbling may occur.

Adhesive performance varies between wood species and some, such as teak, can be troublesome. Where adhesion is a problem it may be possible to add small amounts (~5%) of aliphatic polyisocyanate stabilised with acid functional acrylate. This will boost the curing of material that is inhibited by wood extractives. Another technique that can boost practical adhesion is to temporarily under-cure the priming coat in order to enable a longer period for penetration. This may be achieved by reducing lamp intensity in UV curing, or to reduce the number of lamps.

High primer adhesion may also be attained using epoxy acrylates; however, because these are quite viscous heated rollers may be necessary for application (Table 3).

Table 3 Some examples of technology and their application

Technology example	Typical application
Epoxy acrylate or polyester sealers	Open-pore clear coating
Epoxy acrylate/urethane acrylate topcoat or polyester acrylate	Roller coating + UV cure
Epoxy acrylate + amino polyether acrylate	Three-layer clear coating. Open pore
UV-curable polyurethane dispersion	Spray-applied UV-cured open-pore clear coating (lower gloss)
High solids (100%) UV systems	Clear open-pore coatings Spray or roller applied
Pigmented epoxy acrylate sealer + pigmented urethane acrylate	White opaque hardboard finishes Roller coated
Pigmented UV-curable polyurethane dispersion	White opaque coatings for spraying wood
UV-curing PU emulsion primer Physically drying PU dispersion topcoat	Kitchen cabinets

---

CONTROL OF FILM THICKNESS

Film thickness has a major influence on both appearance (e.g. open/closed-pore effects) and performance. It is partly controlled by the solids content of the coating but also the spreading rate. Roller coating primers will typically be applied in the range 10–25 g m^{-2}. Sprayed water-borne clear coatings can be applied at 60–70 g m^{-2}; thicker coatings may skin and bubble. With curtain coating, coating weights of 80–100 g m^{-2} are achieved and with UV curing it is necessary to balance the photo-initiator level.

Most industrial coating processes will use several layers to build up the desired appearance within operational constraints.

3.7. Coating systems for some typical applications

Different systems for 3D, flat (transparent and opaque) and moulded surfaces (transparent and opaque) are presented in the following illustrative examples:

- Wooden chairs (beech) closed pore, transparent, various gloss levels:
  - Sanding 180 grain
  - Application of WB stain by flow coating

- Drying (2 h) ventilated oven
- Application of 2K-PU by electrostatic or HP air-assisted spray
- Drying (2 h) ventilated oven
- Sanding 250 grain
- Application of 2K-PU by electrostatic or HP air-assisted spray
- Drying (2 h) ventilated oven

Notes. Typical VOC ~200 g per chair, however, with a water-borne poly-urethane this can be brought down to ~30 g per chair. Drying and sanding periods may need to be adjusted.

• Coating system for flat veneered particleboards: transparent, closed pore, various gloss levels:
  - Sanding grain: 120–150–180
  - Application by roller coating – solvent-based stain
  - Drying: 30-min convection oven
  - Application by roller coating – UV acrylic basecoat
  - Partial drying 1 UV lamp HP
  - Application by roller coating – UV acrylic basecoat
  - Drying 3 UV lamp HP
  - Sanding grain: 320–400
  - Application by curtain coating – 2K – acrylic topcoat
  - Drying: ventilated oven, 60 min

Notes. According to the application rates, the amount of VOC involved in such application can be estimated to be around 92 g m^{-2}. To reduce the potential emissions, the use of a photo-curing acrylic topcoat, applied by rollers (30 g m^{-2}), is an option.

Another option is the use of water-borne coating. A two-pack poly-urethane topcoat could represent a valid solution. With a reduction of the application rates, the same coating system can be used in combination with porous wood species to obtain an open-pore effect.

Similar coating systems can be used also for paper-faced panels imitating the wood appearance. Stain application is replaced by surface cleaning (with acetone, or preferably a low solvent cleaner) followed by the application of an adhesion promoting primer.

The edges of such elements can be coated on stacks by spray application of PU or other coating products. An alternative is the application of plastic or laminates pre-finished edges.

• Coating system for flat paper-faced particle (or MDF) boards: closed pore, opaque, variable gloss levels:
  - Cleaning
  - Application by roller coating – PU adhesion primer
  - Drying: IR lamps, 30 s
  - Application by roller coating – UV acrylic basecoat
  - Drying and curing – 3 UV lamps HP

- Sanding grain: 320–400
- Application by curtain coating – UV acrylic topcoat
- Drying and curing: 3 UV lamps HP

Notes. The total VOCs involved is calculated to be 61 g m^{-2}. For best quality, the topcoat (solvent- or water-borne) can be applied by spraying.

Similar coating systems can be applied directly to MDF panels. In this case, it is appropriate to apply a filling priming coat to homogenise the surface.

- Coating system for veneered cabinet doors: transparent, closed pore, different gloss levels:
  - Automatic spray application – solvent-based stain
  - Drying: IR lamps for 30 s
  - Automatic spray application – 2K-PU acrylic basecoat
  - Drying at 30 °C for 60 min
  - Sanding grain: 320–400
  - Application by automatic spray – 2K-PU acrylic topcoat
  - Drying at 30 °C for 60 min

Notes. In this case, roller or curtain coaters cannot be used because of the presence of cavities in the substrate. The total emission is calculated as 335 g m^{-2}.

The internal part of the cabinet doors is generally finished with one single coat.

The use of a water-borne coating system is an alternative to reduce VOC emissions.

- Coating system for MDF cabinet doors: opaque (white), closed pore, various gloss:
  - Automatic spray application – 2K-PU primer
  - Drying at environmental temperature
  - Application by automatic spray PE (polyester) basecoat
  - Drying at 20 °C for 90 min
  - Application by automatic spray PE basecoat
  - Drying at 20 °C for 16 h
  - Sanding grain: 320–400
  - Application by automatic spray – 2K-PU topcoat
  - Drying at 20 °C for 4 h

Notes. The VOC involved in this coating system is calculated to be around 420 g m^{-2}.

An important variable of this calculation depends on styrene, a reactive solvent present in many polyester coatings. This monomer (or reactive solvent) reacts chemically with the film former during the curing process becoming part of the solid coating film. However part of it, due to the relative low vapour pressure, also evaporates together with the other solvents.

The amount of styrene evaporating from the substrate depends on the specific working conditions (temperature, ventilation). In the case presented above, the polyester is considered to be formulated with the 15% of solvents and the 30% of styrene. One third of styrene (10%) is regarded as being released.

Alternatively, a similar coating system could be based on water-borne priming and topcoats. With similar application rates, the VOC involved would be considerably reduced.

It may be optional that the internal surface (not visible) can be coated with a melamine paper, without any further application of coatings.

- Coating system for wood profiles: closed pore, transparent, various gloss levels:
  - Sanding grain: 120 and grain: 180
  - Application by roller (or spray) – solvent-based stain
  - Drying: ventilated oven, 30 s
  - Application by spray – PU basecoat – four passes
  - Drying: air-ventilated oven, 30 s after every pass
  - Brushing
  - Application by spray – 2K-PU topcoat
  - Drying at 20 °C for 24 h.

Notes. The total VOC involved in this coating system is calculated to be around 770 g m^{-2}. This value is too high to meet legislative requirements and unless abatement techniques can be used it will be necessary to adopt a compliant system. The finishing of wood profiles can be carried out with the use of photo-curing products applied with vacuum systems. When 'appearance' plays a determining role, the use of water-borne coatings (e.g. UV) applied by spraying represents a valid alternative to traditional solvent-based coatings. Similar considerations apply to 'old style reproductions' (simulated antique effects) in which solvent-based nitro-cellulose lacquers have been used.

4. COATINGS FOR WOOD FLOORING

Wood flooring utilises the timber of hardwoods and also softwoods such as spruce or pine. It may be installed directly into buildings and finished on site. Finishing will involve sanding as well as coating. This is increasingly a professional rather than DIY process and may use two-pack polyurethane products, or even a portable UV-curing apparatus. Subsequent maintenance in larger buildings such as sports facilities and schools also requires professional expertise. In domestic housing maintenance may be carried out with DIY decorative products.

Hardwood floors may be installed over wooden sub-floors or over a concrete slab to give greater dimensional stability. An alternative to solid wood is to use 'Engineered Hardwood' which comprises layers of hardwood veneers glued with the grain of alternate layers at right angles to each other. They are usually cut to have a tongue and groove. Installed hardwood floors have a pleasing decorative appearance. The general effect overlaps with parquet flooring where sections of wood are laid in a geometric pattern. Many patterns are possible though the herringbone effect is most common. Parquet is made from many different woods including oak, walnut, cherry and maple. Bamboo may also be used (strictly speaking this is a grass rather than wood).

Although wood and parquet can be finished in situ, it is increasingly likely that it is pre-finished in a factory; modern coatings will greatly improve the wear resistance (Fig. 4). Factory coating of wood flooring is thus a significant market sector. Another growth area in flooring has been the use of laminates. This is often made to simulate wood (as well as other materials such as marble or stone), but is actually a facsimile. Laminate flooring evolved from high-pressure melamine laminates widely used for kitchen working tops. The evolution of hard-wearing surfaces combined with packaging in easy to use tongue and grooved strips led to a rapid growth in flooring applications. It has proved highly popular as a DIY product and is usually less expensive than solid wood. Laminated flooring comprises a backing layer supporting a core and a decorative layer. Over this is a 'wear layer' to protect the surface; typically based on melamine

Figure 4   Example of wood flooring (courtesy: Salvador Dalvano).

resin and aluminium oxide. In direct pressure laminates 'DPL' the layers are fused and impregnated simultaneously. The high-pressure laminate process 'HPL' separates the process into two bonding stages. Laminate construction is not a finishing operation comparable to solid and parquet flooring though it does use resins that are also used for coatings. Although flooring laminates are very hard wearing, there will be some requirements for renovation in situ; this is not yet a well-developed market.

4.1. Formulating parquet and related coatings

In common with other coatings, the formulation of a parquet finish requires a clear specification and suitable evaluation methods leading to a specification. Raw materials meeting the functional and operational requirements can then be selected. Clearly, scratching and abrasion are daily occurrence on flooring, and coatings must resist actual damage, but also minimise the appearance of wear (avoidance of 'white lines'). Marking by shoes heels (BHMR = black heel mar resistance) is also undesirable. Binders thus need to combine toughness and hardness and may be reinforced with fine particles increasingly in the nano-size range; both silica and alumina particles are used. UV absorbers may also be used to minimise colour change in areas of differential lighting.

However, hard wearing the floor coating is, the point will be reached where maintenance becomes necessary. This can lead to a difficult stripping operation before renewal. Some companies are therefore developing priming coats which aid stripping, but do not degrade the performance of the wear layer.

4.1.1. Recent developments in parquet coatings

The need for extended lifetimes has led to considerable investigation into products and process that improve performance. Treatment of the wood itself has been one approach. This includes various chemical treatments such as the use of cyclic n-methylol compounds. Plasma treatment has also been investigated as a means to improve adhesion. Most of the modified woods (described in Chapter 2) such as 'thermowood' have also been evaluated as flooring materials, and for their interaction with coating systems.

Developments in binder technology have not surprisingly been driven by VOC considerations, as well as an overall desire to improve performance. Examples include:

- Self-cross-linking water-borne acrylic polymers
- Polyurethane dispersions (PUDs)
- 2K water-borne polyurethane secondary dispersions
- UV-curing aliphatic urethane acrylates

Primers may be formulated differently to aid adhesion and employ epoxy acrylates and aromatic urethane acrylates. Alkoxy functional silicone resins have been shown to improve scratch resistance.

One of the difficulties in evaluating these developments is to find tests methods that correlate with practical experience. Alternative test methods have been found to rank coatings in different orders; this is a familiar problem in accelerated testing! The Taber abraser (Chapter 6 – ISO 7784–2) is widely used as a general test method but has not given good correlation with real damage patterns. The correlation is generally worse when coatings are textured, as is the case with some non-slip surfaces. Another abrasion device is the Martindale apparatus which imposes a pattern of scratches known as 'Lissajous' figures has also been investigated as part of an ongoing project on the wear resistance of flooring [4]. There are many other mechanical tests which relate to performance but not in a direct way, and this is the subject of ongoing work. Micro-indentation methods can be used to investigate 'residual creep' [5] which might be related to energy absorbed during damage. The motor car industry has made progress in designing coatings that recover (self-healing) after damage by, for example, the actions of a car wash. These materials are based on 2K polyurethane resins and it may be expected that this technology will have applications in flooring. However, the requirements are different and floor coatings must also sustain loads over long periods without blocking.

4.2. Industrial application of parquet coatings

Floor coatings are typically applied by roller application; this is also the favoured method for imparting a textured anti-slip surface (Fig. 5).

Phase 1. Substrate sanding

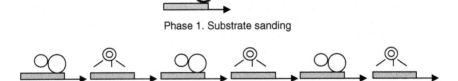

Phase 2. Primer application (UA) with roller coaters (20 g/m2 per coat) + drying with 1 UV lamp ovens (80 – 120 W/cm).

Phase 3. Sanding + UV Top coat application with roller coaters (8 - 15 g/m2) + drying with a 3 lamp oven (80 – 120 W/cm).

Figure 5 Parquet coating process.

A typical coating schedule would be included sanding followed by two coats of UV-curing acrylic basecoat (spreading rate 30 g m^{-2}), the first coat would only be partially cured ('jellification'). After sanding with a fine gain abrasive two further coats of UV acrylic topcoat are applied (spreading rate 10 g m^{-2}) with only partial curing between coats. Since the transfer efficiency is very high, and the VOC low, the overall VOC emissions are very low. Although the whole system can be applied using rollers a more uniform surface can be obtained with curtain coating for the topcoats; however, some thinning of the material may be necessary and VOCs will be higher.

The different levels of curing between coats are achieved by adjusting the number of lamps (e.g. UV2000) from one to three. In practice, many different plant configurations are possible with additional stages for special effects and other finishing operations. An account of a modern coating plant for parquet flooring is given in Ballardinin [6].

UV lacquers can also be applied in situ using portable apparatus. There have also been developments in the use of photo-initiators which are activated by sunlight (fairly lengthy) or by fluorescent tubes (2 h). Photo-bleachable dyes provide a marker for the correct dosage.

Although UV curing is now the most common method for factory coating of parquet flooring, traditional oil-based finishes find a role where a traditional more open-grained effect is required.

REFERENCES

[1] James, M. (2008). Film biocidal active substance with an environmentally favourable profile. *In* "Proceedings of the 29th FATIPEC Congress", Ghent, Belgium(paper 84).
[2] Mestach, D., Akkerman, J., and Sanderse, A. (2008). New fast drying waterborne two pack urethane coatings for industrial wood finishes. *In* "Proceedings of the PRA 6th International Woodcoatings Congress", Amsterdam (paper 15).
[3] Schipholt, N. L., and Beckers, E. P. J. (2008). Factors influencing wet look "Anfeuerung" of clear coatings on wood. *In* "Proceedings of the PRA 6th International Woodcoatings Congress", Amsterdam (paper 28).
[4] Emmler, R., and Nothhelfer-Richter, R. (2005). Scratch-proofing: Towards a standard scratch resistance test for parquet coatings. *Eur. Coatings J.* **1–2**, 36–40.
[5] Vu, C., Ferte La, O., and Eranian, A. (2005). High-performance ultraviolet multilayer coatings using inorganic nanoparticles. *In* "Proceedings of the RadTech Europe 05", Barcelona (paper 8).
[6] Ballardinin, R. (2006). Choice of a coating plant for a refined and versatile parquet floor. *Verniciatura del Legno* **147**, 78–83 (English), 29–35 (Italian).

General information deriving from European Coatings Conference: Parquet Coatings IV", 9–10 November 2006, Vincentz Network GmbH, Germany.

BIBLIOGRAPHY

Bulian, F., and Tiberio, M. (2002). "Distretto della Sedia: Gli aspetti ambientali e le migliori tecnologie", vols. 1 and 2, CATAS.

Bulian, F., and Tiberio, M. (2008). La riduzione delle emissioni di composti organici volatili nel settore legno-arredo, Federlegno.

Bulian, F. (2008). "Verniciare il legno." Hoepli, Milano.

CHAPTER 9

Operational Aspects of Wood Coatings: Application and Surface Preparation

Contents			
	1.	Introduction	259
		1.1. Quality	260
		1.2. Application and spreading rates	260
		1.3. Productivity	261
		1.4. Cleaning/product change/maintenance	261
		1.5. Transfer efficiency	261
	2.	Application Systems	262
		2.1. Contact methods	263
		2.2. Atomising systems	273
	3.	Preparation of the Substrate	283
		3.1. Sanding of the substrate	283
		3.2. Sanding papers	284
		3.3. Brush sanding	285
		3.4. Bleaching	287

1. INTRODUCTION

Successful wood coating involves several stages including preparation of surfaces followed by the application, and drying of coatings. Collectively, these activities may be described as 'operational aspects' and are the subject of this and the next chapter. Broadly speaking, these issues relate primarily to the industrial wood coating sectors (joinery, flooring and furniture) and in a more limited way to the decorative/architectural sector.

Wood Coatings: Theory and Practice
DOI: 10.1016/B978-0-444-52840-7.00009-6

'Application' refers to a process which coating material is transformed from the bulk state in containers, to a more or less even and smooth film on a prepared substrate. Thereafter various drying and curing stages will occur with optional finishing operations, such as sanding between coats. Application processes are selected on the basis of speed, suitability, material, cost, etc. Whereas hand application may be suitable for house decorating and small-scale operations, automation and robotics are likely to be involved for many industrial applicators. The operational criteria that may be used to appraise application processes include:

- Quality of finish
- Application rate
- Productivity
- Maintenance and turn-around time
- Transfer efficiency
- Technical capability
- Economics

1.1. Quality

The quality of coated surface can be considered in terms of appearance and performance. Uniformity in the coating layer and good 'levelling' strongly depend on the application system used, and on its proper set-up. They will also be modified by the properties of the coating itself such as viscosity and surface tension. Surface appearance will determine the impact of the finished product towards the final customer and is fundamental for its success.

Application can also affect the final performance of the coated product. The inclusion of air bubbles, for example, will not only impair appearance but also compromises film integrity and resistance properties. Poor wetting of the surface can lead to adhesion problems.

1.2. Application and spreading rates

The application rate for a given process is the amount of a coating required to produce, under defined working conditions, a dry film of specified thickness; it is therefore expressed in volume or weight terms per unit area (e.g. $l\ m^{-2}$ or $kg\ m^{-2}$). Application rate is mainly dependant on the transfer capability of the equipment and the viscosity and solids content of the coating. The application rate of a given system is usually expressed with a 'range' including the minimum and the maximum values.

The reciprocal of the application rate is called spreading rate. It expresses the surface area, which should be covered by a given quantity of coating material to produce a coating film of the required thickness ($m^2\ l^{-1}$ or $m^2\ kg^{-1}$). In selecting an application process must be capable of an application rate that will meet the specified spreading rate.

1.3. Productivity

Productivity is a criteria used to quantify the throughput of an operation. It can take many forms, the simplest of which is expressed by the total square metres of substrate coated per minute. However in some circumstances, it may be more relevant to consider the number of items coated and to take into account reject or repair rates. Productivity is a fundamental parameter for industrial production with a major impact on process economics.

1.4. Cleaning/product change/maintenance

The cleaning of an application system is an important industrial operation because it determines the economics of passing more or less easily and quickly to the application of another product having, for example, a different colour. As such it contributes to overall productivity. Waste materials produced during cleaning, such as contaminated solvent, represent also an additional cost and may be penalised by solvent management legislation.

Equipment maintenance is another economic issue that reduces productivity if equipment is taken out of service. The cost of spares and their sourcing must also be taken into account.

1.5. Transfer efficiency

Transfer efficiency (TE) is a value, expressed as a percentage, representing the ratio of the amount of coating applied onto the substrate to the total amount of coating material consumed for that application.

The greater is the transfer efficiency, the lower is the relative coating consumption; conversely low-transfer efficiency means high wastage. Apart from the economical benefits deriving by the use of high-transfer efficiency systems, a lower emission of VOC into the atmosphere has positive benefits on the environment and is a requirement for some legislation. Coating material not deposited on the substrate by a spraying process is called overspray; under some circumstances, this may be recovered and thus improve overall efficiency:

$$TE\,(\%) = 100 \times (Q_a/Q_t),$$

where TE is the transfer efficiency, Q_a is the amount of coating material applied onto the substrate and Q_t is the total amount of coating material used.

Where recovery systems are used, the overspray can be totally or partially recycled. In these cases, the calculation of the transfer efficiency of the overall system should take into account the recovered product.

Table 1 Transfer efficiency of different application systems

Application method	Transfer efficiency (%)
Roller coaters	95–100
Curtain coaters	95–100
Vacuum applicators	95–100
Pneumatic spray guns	30–45
Airless spray guns	35–50
Electrostatic-assisted spray guns	40–60
Electrostatic-assisted rotating systems	80–90
Powder spraying	95–100

When changing processes, it is often useful to calculate any improvements in transfer efficiency which may be useful in meeting VOC emission requirements. The following equation expresses the reduction in the coating consumption as percentage.

The calculation is valid if all the other parameters (e.g. coating dilution) remain the same; otherwise, a correction for solids content should be made:

$$\text{Consumption reduction } (\%) = 100 \times \frac{\text{TE}(\%)_{\text{after}} - \text{TE}(\%)_{\text{before}}}{\text{TE}(\%)_{\text{after}}}.$$

In Table 1, the transfer efficiency of various application methods is listed. These data are taken from various publications and will be affected by the procedures followed for their calculation (specific application system, adjustment, coating material used, environmental conditions, etc.). The data can be considered as examples of transfer efficiency values achievable with typical equipment. They cannot be considered as absolute values.

2. APPLICATION SYSTEMS

The process of application is different for powder, as opposed to liquid systems. In the case of liquid coatings, a distinction may be made between:

- *contact methods*, where the common principle is the continuous contact among the three elements: application system/coating/substrate; and
- *atomising methods*, where the liquid coating is converted into small droplets and launched towards the substrate.

A further distinction can be made between manual and automated methods.

2.1. Contact methods

2.1.1. Brushing

Brushing is the simplest application system presenting some advantages like: low cost, high-transfer efficiency and low capital cost. It has been established for hundreds of years and is widely used for architectural painting. For industrial purposes, the low productivity (typically around 1 m^2 min^{-1}) and the relatively low quality of the final result limit the applicability. However, brushing may be useful for touch-up after sanding operations have partially removed the colour imparted to wood by tinting or impregnating products. Mohair, fabric, sponge and other simple rollers may also be thought of as a kind of rotating brush (as opposed to the industrial roller coating discussed below). They also show good transfer efficiency and are faster than brushes for large areas. Rollers can be fitted with various feed and pressure systems to make the process even faster.

2.1.2. Padding

Padding is another simple manual system still used by some craftsmen. Special plugs made of fibres or textiles are used to spread on the substrate semi-solid coating products like waxes or shellac. The repetition of such operations allows a deep penetration of the coating into wood pores and a homogeneous application result. Variants of pads are also found in the decorative market and may even be pressure fed for the quick application of fence treatments.

2.1.3. Dipping

Dipping is a practical, economic and fast application system mostly used with low viscosity coatings including stains and impregnating stains. It is suitable for elements presenting complex shapes with difficult accessible sides. Dipping can also be used for the application of non-filming coatings and, in some cases, for the finishing of low costs products like, for example, beech chairs for which the required quality is relatively low. A poor finish may result if the liquid coating runs and sags; automated withdrawal will reduce this. Dipping systems are suitable in particular for water-based products as the potential pollution and risk of evaporation from open tanks is less of a problem than is the case with solvent-based coatings.

2.1.4. Autoclave vacuum and pressure application

Autoclave and other closed vessel processes are particularly used for wooden products to be exposed outdoors. The main benefits of such processes are deeper impregnation of wood than can be achieved by simple dipping. When used with appropriate preservatives, this enables better resistance against biological deterioration as a consequence of the more effective impregnation. Nonetheless, it should be noted that

European Standards have moved to performance-based, rather than process-based criteria. Performance will depend on additional factors such as efficacy, retention time and leaching rates. There are several variations of the impregnating process according to whether or not vacuum is applied and to the pressure cycles used. The following steps are typical of the 'double-vacuum' process:

(1) Pressure is reduced inside the autoclave to suck the air from the wood inter-cellular voids.
(2) An impregnating treatment or stain is introduced in the autoclave. Penetration into wood interstices is promoted by the pressure difference so generated.
(3) Pressure inside the autoclave is raised to the normal atmospheric values.
(4) The surplus liquid is removed from the system.
(5) A second depressurisation is used to remove any excess of impregnating stain from wood.
(6) Pressure is returned to atmospheric and the process is completed.

2.1.5. Roller coating

The basic principle of industrial roller coating is the initial distribution of the liquid onto the surface of a suitable roll. By the simple rotation of the roll directly in contact with the substrate, the coating is transferred on its surface. Rollers also transport the substrate and in some equipment support the substrate against a doctor blade or another roller that applies the coating.

The requirement for a fixed gap at the point of contact makes roller application possible only on completely flat surfaces. Three-dimensional elements or moulded panels are not suitable for such systems and even with flat surfaces the coating of edges may be carried out with other methods (e.g. spray application).

Roller coaters are used for the application of stains (solvent or water based) and one-component coatings (primers, priming coats and topcoats). They are particularly useful for the application of radiation curing products, as their high viscosities and solid content allow the achievement of good results in terms of film thickness and surface homogeneity. As roller coaters provide the recovery and recirculation of the coating, they are not really suitable for the application of two-pack systems which have a limited pot life.

The main advantages of roller coating can be summarised as follows:

- The ability to regulate the application rate from few grams per square metre up to 50 g m^{-2} and more in the case of multi-roll systems
- Good productivity as the systems can be completely automatic
- Capability to use high solid content coatings (e.g. radiation curing) due to their suitability for high viscosity products

- Recovery and recirculation of the product not coated. The transfer efficiency can be close to 100% for a long run of a single product

There are many types of industrial equipment utilising rollers. These machines are equipped with of one or two heads each consisting of an applicator roll, which applies the coating onto the substrate and a feeding (or dosing) roll, which proportions the coating to be applied. In many configurations, the dosage rate is controlled by a 'knife' which may itself be fed by a roller. When the substrate and roller travel in the same direction and usually at the same speed, the method is called 'direct roller coating'. On the other hand, when the roller motion is opposite to the surface movement, of the substrate the method is called 'reverse roller coating'. The roller coaters differ substantially in the number of rollers, in their rotation direction and for the material covering the rollers (steel or rubber) (Fig. 1).

An oscillating scraper (frequently called the 'doctor blade') is provided to keep the application roll cleaned and to remove any coating not applied to the substrate.

Flat sheets, for example, MDF and other substrates, are transported by conveyor belts or by small parallel rollers (steel or rubber coated).

2.1.5.1. One-head roller coaters   One-head roller coaters are equipped with two supplementary rollers (feeding and applicator). The rotation direction of the applicator roll is in the same direction as the panel feed. The two rolls can be direct or reverse. For some coaters, it is also possible to independently control the movement speed of the panels. One-head roller coaters are used for the application of stains, primers, priming coats and topcoats.

Very often, the coating systems used for the application of stains represents a special category of one-head roller coaters known as 'staining roller coaters'.

The coating material flows between the two rolls. The application rate is determined primarily by the gap between the rolls, their speed and the pressure exerted against the substrate.

In the case of direct roller rotation, the amount of applied product is higher, but the result is coarser, and more suitable for the application of priming coats.

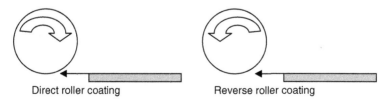

Direct roller coating   Reverse roller coating

Figure 1   Direct roller coating and reverse roller coating.

Figure 2   One-head roller coater (courtesy: Superfici spa).

For the application of topcoats, the rotation direction of the two rolls is reversed. In this way, the feeding roll tends to stretch and distribute more uniformly the coating on the applicator roll, obtaining a more homogeneous film onto the substrate. Quality of application can be further improved also by using a belt conveyor, with speed adjustment (Fig. 2).

Application rates of such systems can vary from 10 g m^{-2} for the topcoats up to 30 g m^{-2} for the priming coats. Application rollers are usually coated with rubber, the composition of which depends on:

- Application rate
- Chemical resistance to any solvents present in the coating materials
- Hardness

To maintain the roll in a clean manner, free from accumulations, the main roller is fitted with a metal scraper blade running across the full length.

2.1.5.2. Staining roller coaters   Staining roller coaters are the simplest roller systems used for the application on flat surfaces of stains or primers. In the case of stains, roller coaters are usually equipped with wiping rolls cleaning the substrate before the application. The dosing rolls made of chromium-plated steel, and the application roll, constitutes the head. The surface of the head is coated with a soft rubber (e.g. 30 shore) in the case of solvent-based stains; or with a sponge material if water-borne stains are applied. The application rate typically varies between 10 and 50 g m^{-2}.

One or two brushing (or wiping) rollers can follow the application head.

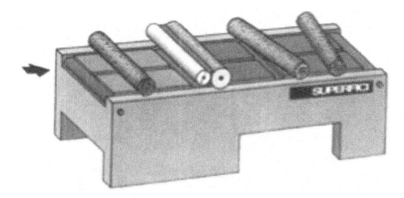

Figure 3   Staining roller coater (courtesy: Superfici spa).

Staining roller coaters are particularly utilised with wood species that have large pores (e.g. oak). Their uniform application allows good penetration of the stain into the pores and removes the unwanted excess (Fig. 3).

A further sub-category of staining roller coaters are the printing rollers. They are able to reproduce the grain of any wood species onto the surface of wood-based panels covered with ordinary veneers or onto other surfaces like PVC foils or a suitable painted basecoat. Printing rollers are normally equipped with one head consisting of two rollers (inking and off-set). The metal surface of the first is engraved with the grain pattern to be reproduced. The off-set roller, rubbery coated, receives the ink from the first and transfers the pattern directly onto the panel surface. Printing is usually carried out by some successive applications, onto a slight coloured priming coat also applied by roller coating.

Printing roller coaters presently used mainly utilise photo-curing inks. Grain printing is an economic way of reproducing expensive woods. High precision of the latest techniques and great productivities are valued attributes of such systems.

2.1.5.3. Two-head roller coaters (dual or twin head)   Application with one-head coaters can sometimes lead to the formation of slight longitudinal stripes due to the difficulty of maintaining a constant and homogeneous coating layer on the rolls. Subsequently sanding is more difficult, leading to negative effects on surface appearance. One solution to this problem is to decrease the application rate or reduce the viscosity of the coating; however productivity is reduced. Roller coaters equipped with two heads allow a double application which improves the homogeneity while maintaining an adequate application rate. The application rate of the second roll is usually lower than the first but the double application

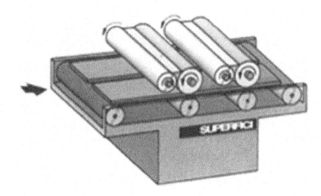

Figure 4   Two-head roller coater (courtesy: Superfici spa).

gives a more homogeneous result. A typical set-up will use a rubber application roller followed by a smooth chrome steel roller, but many configurations are possible (Fig. 4).

2.1.5.4. Filling roller coaters   Filling roller coaters are equipped with three rollers (dosing, applicator and smoothing). They are similar to the one-head coaters, with the doctor and the application rolls in reverse rotation (Fig. 5).

The third roll is also made of chromium-plated steel, rotating in the opposite direction to the conveyor belt.

The smoothing roll has two main functions:

(1) Firstly to remove any excess of coating applied
(2) Secondly to achieve a uniform applied layer by forcing the coating inside the pores and cavities of the substrate

In the UK this approach is sometimes described as an 'Italian finish'.

By varying both the pressure against the substrate and the speed, it is possible to accurately regulate the amount of coating applied. With filling

Figure 5   Filling roller coater (courtesy: Superfici spa).

roller coaters is possible to apply considerable amounts of fillers or priming coats usually in the range of 30–40 g m^{-2} for porous substrates (e.g. particleboards) and 20 g m^{-2} for more compact surfaces such as MDF. As the smoothing roller is frequently independent from the coater head, it can be manually lifted with a hand wheel enabling the filling roller to also be used as a simple one-head coater.

2.1.5.5. Reverse roller coaters   The name derives from the rotation direction of the applicator roll, which is being in the opposite direction to that of the panel (see Fig. 6).

As a consequence of the reversal, the applicator roll is better able to smooth and accurately dose the applied coating.

These coaters can be divided into two categories:

(1) Two rollers coaters (called 'baby reverse' or simply 'baby')
(2) Four rollers coaters (simply called 'reverse')

With the 'baby' coaters, the application rate can reach values around 30 g m^{-2} which is usually suitable for UV topcoats (clear or opaque).

The four-roll system is a combination of two heads the second of which is the reverse roll. By the de-selection of the second head, the system can be used as a simple one-head roller coater. With four-roll coaters a double application can be obtained allowing an application rate in a wide range of values from 10 up to 100 g m^{-2} and more. Various typologies of priming coats and topcoats can be applied with high-quality results. The application rate depends on many factors:

• Rotation speed
• Gap between the applicators and the feeding rolls
• Pressure
• Rubber hardness
• Conveyor speed

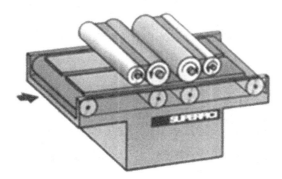

Figure 6   Reverse roller coater – four roller (courtesy: Superfici spa).

2.1.6. Curtain coating

Curtain coaters are application systems suitable for flat surfaces. They utilise the capability of some coatings to form a uniform thin liquid layer flowing from a narrow slot or over a blade. A typical configuration comprises of one or two heads and a conveyor belt for the transport of the panels. Narrow tanks (or weirs) fitted with slots in the bottom comprise each head. After passing through the slot, the flowing coating material forms a thin stable and uniform curtain. The coating flows are driven either by gravity alone or with pressure assistance. A recovery system, placed below the head, provides options for recycling. As the panel passes through the curtain, it is directly transferred onto the panel surface.

Curtain coaters are fast and efficient systems operating at up to 100 m min^{-1}. Although flexible enough to be used for the application of various coating types, the most common utilisation is for topcoats.

They are suitable for both water-borne and solvent-borne polyester or acrylate coatings. Because curtain coating relies upon recirculation, volatile compounds can easily evaporate from the curtain and viscosity need continual adjustment.

Spreading rate depends mainly on the conveyor speed; typical values are between 100 and 300 g m^{-2} and sometimes more (Fig. 7).

The case of two-pack materials with a limited pot life, such as chemically initiated polyesters, would be problematic for a single-head coater. However, the use of a second head can overcome this difficulty. With the first head the polyester mixed with the peroxide is applied; the second head, placed just beyond the first, applies the same polyester mixed with the accelerator (a metal salt). The curing reaction takes place between the two layers directly on the substrate. Another variant for two-pack materials is to combine a roller and curtain coater in sequence.

The limits of curtain coatings can be summarised as follows:

- Complex cleaning and maintenance
- Possibility of curtain breaking in use due to simple air draughts
- Suitability only for horizontal plane surfaces. The presence of profiles and indents could cause uneven film thickness to occur

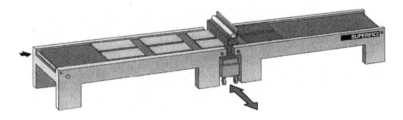

Figure 7   Curtain coater (courtesy: Superfici spa).

A practical concern concerning the panel, coated by curtain or roller coating, is presented by edges. Edges are not properly coated by roller or curtain coaters and it is often considered better to protect the edge altogether. Adjustable magnetic bars are sometime used to provide such edge protection. The coating of the edges must then be carried out by other means either prior to, or after the main face has been coated. Typical solutions are:

• Application of paper or plastic profiles (PVC, ABS)
• Use of pre-coated wood profiles
• Spray coating of the edges with several stacked panels (e.g. stain, priming coat and topcoat)

2.1.7. Flow coating

Flow coating is a procedure in which coating materials are pumped or flooded over components to be coated. It thus represents an alternative to simple dipping. In some systems, the coating materials are sprayed at low pressure with coarse nozzles to components hanging in overhead conveyors. The coating is directed in coarse droplets to flood the surface while excess material flows to the bottom of the application chamber for recycling (Fig. 8).

The method is thus considered a contact rather than atomisation process as described below. The main advantage of flow coating systems is high productivity.

Flow coating systems are particularly suitable for low viscosity coatings such as stains, impregnating stains and intermediate coatings. Three-dimensional elements (e.g. component or already fully assembled window frames or chairs) can be completely coated by such systems. For the same environmental reasons reported for dipping systems, VOC emissions are a concern and flow coating technology is being adapted for water-based products. A modification of flow coating uses horizontal line feeding of profiled pieces ('line flow type'), and has been introduced for the application of water-based products on disassembled frame parts.

2.1.8. Vacuum coaters

Automatic vacuum coaters are generally used for the continuous all round finishing of long and narrow elements either flat or profiled, in combination with UV curing coatings. The closed application cell, inside which the pressure is strongly reduced, represents the 'heart' of such systems. Vacuum coaters require low viscosity coatings that are metered by a profiled template at the inlet and outlet side of the cell with a tolerance of about 3 mm. The gap may be adjusted to control film thickness and the vacuum removes excess material prior to curing.

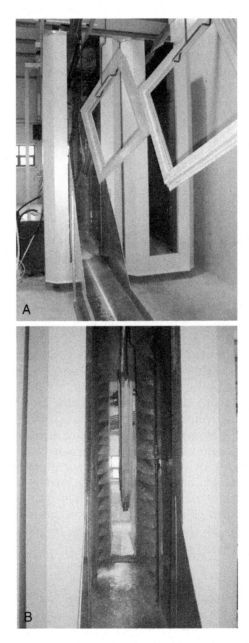

Figure 8   Flow coating systems for window frames (courtesy: Vertek Group srl).

The main advantages of such systems are represented by the high productivity deriving from the simultaneous application on all sides of the profile and very fast line speed. However, maintenance of the vacuum pumps is critical and they have a high-energy requirement.

2.2. Atomising systems

2.2.1. Conventional air-atomised systems – Pneumatic atomisation

Pneumatic atomising systems are commonly known by the collective term of 'spray application' a process invented by Dr De Vilbiss in 1890 and still a well-known company. Spraying is a versatile process, which can be manual or automatic and used with most coating types. It has the great advantage of laying down a smooth film applicable to intricate shapes. However, a major drawback is dealing with overspray to obtain good transfer efficiency. The principle of spray systems is to atomise the coating materials by means of suitable 'guns'.

The gun consists of a body, a head and a nozzle. Guns are able to reduce the coating materials into small drops directing them towards the substrate. Pneumatic, hydraulic and electrostatic methods are utilised in the many variants of spraying, which are described below. Other possibilities for atomisation are centrifugal and ultrasonic forces described later.

In the case of two-pack coatings, the mixing of the two components can be obtained directly inside the head thus avoiding the problems of a limited pot life. Pneumatic guns by definition utilise air for atomisation, and often also for transport of the coating to be sprayed. The coating is forced through a suitable nozzle and immediately subjected to air pressure by peripheral jets. Air pressure is derived from integral or remote compressors.

The liquid coating flows to the head going out from the nozzle by gravity, or by the suction effect generated by airflow. The coating container can optionally be maintained at a slight overpressure (Fig. 9).

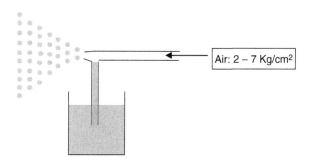

Figure 9   Schematic representation of a pneumatic gun.

The factors influencing atomisation include:

- *Gun configuration*: including needle size, fluid tip orifice and air cap
- *Pressure*: higher pressure gives greater atomisation (smaller droplets). The velocity of the jet induces the effect
- *Surface tension*: reducing surface tension will lower droplet size; however, surface tension gradients can cause defects such as 'orange peel'
- *Viscosity*: like high surface tension, a high coating viscosity acts against the formation of drops. Correct viscosity adjustment (dilution) is a key factor in spray application
- *Temperature*: it has a major effect on atomisation by lowering viscosity and surface tension. Higher temperature will also promote volatisation and cross-linking. This heating can reduce the amount of thinner (diluent) needed and some spray systems are equipped with heaters
- *Density*: for a given set of conditions, higher density gives larger droplets

Pneumatic guns are simple and economic systems for the application of many types of coatings. They usually require low viscosities, providing a good uniformity in the application and in the general quality of the coating film. The main disadvantages are bounce back leading to low-transfer efficiency and in consequence conventional air-atomised systems become unproductive and subject to environmental regulations. The gun set-up (air pressure, coating flow and nozzle) can be adjusted depending on type of coating used, required quality and productivity. Spraying is carried by means of compressed air at a pressure of 2–7 kg cm^{-2}.

2.2.2. Pneumatic atomisation with high air volume and low pressure (HVLP)

HVLP systems have evolved to reduce bounce back and thus improve transfer efficiency from a typical figure of 30–45% for conventional spraying, up to the 60–75% range.

The acronym for these pneumatic systems shows that the HVLPs utilise high air volumes at low pressure. There are several variants some with integral pumps (Fig. 10). However, they can be powered by external compressed air systems through an internal venturi enabling the air cap to operate at very low pressures around 0.7 kg cm^{-2}. Atomisation is obtained by the use of special nozzles. Higher application rates than conventional spray guns are possible, but there is a need for different operator techniques including working closer to the substrate. This requirement may limit the use of HVLP systems in the wood and furniture sector because of the typical presence of wide surfaces or complex three-dimensional objects.

Figure 10 HVLP spray gun for high-transfer efficiency, suited for wood finishing applications (picture is a Devilbiss MSHTE).

2.2.3. Hydraulic atomisation (airless)

In airless spraying, high pressures are used to force the lacquer or paint through nozzles at a rate sufficiently fast to cause atomisation without supplementary air. The lack of moving air improves airless-spraying penetration into areas that would be inaccessible to conventional spray such as the interior large closed containers and where very high application volumes are needed. Water-borne paints require different nozzles from solvent-borne. Equipment is more expensive with additional pumps needed, the high pressure involved require greater attention to safety measures. A potential disadvantage of airless spray in furniture is that high volumes of material are so rapidly applied that control of coating thickness is lost (Fig. 11).

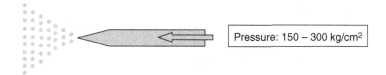

Pressure: 150 – 300 kg/cm²

Figure 11 Schematic representation of an airless gun.

The advantages of the airless application can be summarised by the following list:

- Capability to apply high thickness layers in one single pass
- Suitable for basecoat application on flat surfaces without any cavities or moulds
- High productivity
- Suitable for relatively viscous liquids (e.g. water-borne coatings)
- Transfer efficiency is higher than pneumatic systems

The recognised disadvantages of such systems are:

- The dimensions of the atomised drops are generally greater than those generated by pneumatic systems, leading to a worst result in terms of appearance
- Possible accumulation of coatings in cavities, angles, etc.
- Hydraulic atomising guns are generally more expensive than pneumatic systems
- The high application rate makes the production of low thicknesses layers more difficult, for example, to achieve open-pore effects

2.2.4. Hydraulic air-assisted atomisation

The paradoxically named 'air-assisted airless' is a combination system, which limits the problems noted above by using lower paint pressure, supplemented by a low air pressure to give a more gentle and softer fan pattern. Air assistance is provided by means of air jets positioned on the nozzle head, with a feeding pressure around 40–80 kg cm^{-2}. The airflow helps both atomisation and fan shaping thus allowing the use of a reduced atomising pressure and hence less bounce back (Fig. 12).

The advantages of the hydraulic air-assisted guns reflect those described for the airless systems, but in this case the amount of coating applied can be more accurately controlled. The coating film produced is generally more homogeneous and the appearance is usually better. With these systems, also small elements or complex substrates (three-dimensional, moulded) can be properly finished. This technology is suitable for any kind of coating material (stains, priming coats, basecoat and topcoats).

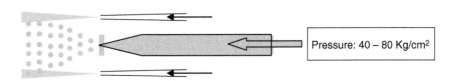

Pressure: 40 – 80 Kg/cm^2

Figure 12 Schematic representation of an air-assisted airless gun.

2.2.5. Operational aspects of spray application

Spray application can be purely manual, or assisted by varied degrees of automation including robotics. The choice will depend on many factors including purchase price in relation to turnover and legislative impact. At the shop floor level, ease of use, skill levels and maintenance must be taken into account.

2.2.5.1. Manual application Operators should operate inside special booths provided with an adequate air circulation. Volatile solvents and particle overspray must be continuously removed from the working area by means of water curtains or dry filters.

Application should always be carried out in the same direction as the air suction flow. Suction intensity and direction will affect application by imparting an additional force on the coating drops.

Manual spraying is applicable when complex shape elements or small lots are to be finished. The quality of the application depends strongly on the operator's skills and training.

2.2.5.2. Robotic application For three-dimensional elements, such as assembled furniture, the throughput may justify spray application by guns mounted on robotic arms. Robots are characterised by the number of movement axes possible. In the simplest systems, the gun traverses only back–forward and left–right. Such systems are particularly used for the finishing of window frames. For more complex 3D systems with cavities and hidden surfaces (e.g. chairs) 'anthropomorphic' robots are used. These mimic arm movements using arms and swivel joints that are computer controlled (Fig. 13).

Figure 13   Spray robotic application (courtesy: CMA Robotics spa).

2.2.5.3. Automatic application   The need for high productivity together with the need for application consistency has led to the development of fully automatic spraying systems. Such configurations are able to apply coatings to flat pieces passing through the spraying area. Application is carried out by means of guns, which move along controlled paths, generally perpendicular, to the forward feeding direction of the pieces. The way the guns are moved is a characterising element of automatic spray systems. The productivity of automatic spraying equipment is dependent on several parameters (Fig. 14):

- Appearance required (e.g. closed or open pore)
- Shape of the work piece
- Gun movement
- Number of guns used

Automatic spray set-ups include several characteristics units which will differ between manufacturers. Components include:

- Conveying
- Automatic cleaning and colour change
- Coating recovery
- Air suction
- Pressurisation
- Waste water treatment (for machines with a water curtain)
- Automatic controls
- Work piece recognition system and gun control electronic station
- Gun movement system

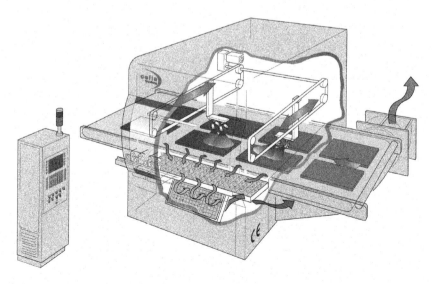

Figure 14   Automatic spray application system (courtesy: CEFLA scrl).

Automatic scanning of work piece dimensions enables reduced coating consumption through more specific and accurate application.

The pressurisation system consists of a fan drawing clean air into the spraying chamber; thus, pollution by overspray or powder present in the air is prevented to alter the quality of the coated surface.

One of the advantages of automatic spray systems, in comparison with roller or curtain coaters, is the ability to coat edges and moulded parts.

2.2.5.4. Electrostatic application   Conventional electrostatic application is a combination of a spray system with an electric generator or power pack coupled with a transformer that imparts an electric charge, directly to the coating (internal charge) or to the particles being produced by the spray system. In the latter case, an electrode is placed adjacent to the exit of the gun nozzle. Ransberg developed the original process in the 1930s, but there are now many variants. All atomising methods can benefit from electrostatic charging and the strong attraction to the target substrate greatly improves transfer efficiency.

An electrode, charged at very high voltages, produces ion species from a point or edge in its surroundings (the corona effect) able to charge the coatings particles emitted through the nozzle. Other modes of charging are contact, induction and friction methods, but contact and corona are most common for liquid coatings. Charged coating particles are directed towards the substrate by the kinetic energy of the sprayed jet and electro-statically attracted to the earthed substrate. This dual effect produces a 'wrap-around' deposition of the coating on 'hidden' surfaces as well as the face.

Factors affecting the electrostatic attraction can be summarised as follows:

• *Substrate conductivity and shape.* The substrate should be sufficiently conductive to discharge the electric charges transported by the coating particles. Dry wood is an insulating material and its conductivity is determined by the humidity content. If the wood elements to be coated are relatively dry, electrostatic attraction would be to other parts of the booth connected to earth rather than the target substrate. To avoid this problem, electrostatic spraying should be carried out in a 'fog chamber' when conditions are dry. Another approach is to pre-coat the wood with a conductive primer. From the perspective of shape, the 'Faraday Cage' effect limits the depth of electrostatic penetration into deep cavities; also rounded shapes are favoured rather than sharp edges.
• *Electric field charge.* The intensity of electrostatic forces depends on two parameters according to Coulomb's law:
  – Charge transferred onto the coating particles (electrode polarisation)
  – Distance between gun and substrate

A typical charge regime is 60–85 kV at up to 150 μA.

- *Coating resistivity.* Resistivity is the degree to which a liquid coating resists the flow of current and conductivity is the opposite. The preferred resistivity range for electrostatic spraying is between 25 and 100 MΩ cm. Some solvent-borne products are considerably above this figure and once resistivity is above 200 MΩ cm electrostatic effectiveness will decrease. However, conductive polar solvents may be added to reduce resistance. With water-borne coatings low resistivity becomes a problem. Electrostatic spray can only be used if the electrically conductive water-borne paint is isolated from the electrostatic system. A number of isolation methods can be used to avoid grounding out the electrostatics in a water-borne system. Mechanical atomisation also offers advantages. Another alternative to isolation is to use a voltage blocking device between the spray head and the grounded paint supply.
- *Particle dimension and speed.* The electrostatic effect is more efficient with smaller particles, as the ratio of charge per unit area is greater. As previously reported, the particle dimensions depend on various factors (spray gun, surface tension, viscosity, etc.). Also the velocity of the coating particles plays an important role. With slower particles, there is more chance that the initial trajectory will change.
- *Environmental conditions.* The evaporation of solvents as temperature increases causes a reduction in the particle dimensions together with an increase of the charge intensity per surface unit. Air humidity is also a fundamental parameter as it promotes air conductibility. The transport of the particles towards the substrate is then promoted. The importance of moisture content to spray wood electrostatically was noted earlier.

2.2.6. Mechanical atomisation

2.2.6.1. Centrifuge atomisation (rotation systems)   Mechanical systems utilise centrifugal force to convert liquid coating materials into small droplets, and may also be used with powders. The most common configuration of centrifugal systems is either a 'bell' or a 'disc' axially rotating at very high speeds.

Rotational application is usually combined with electrostatic transfer as part of an automated production line. Rotational systems show high productivity with good transfer efficiency; however, the positioning of the material may not be as 'accurate' as with traditional spray application. They are mainly used for the application of basecoats:

- *Bell atomisers.* With bell atomisers, the cone shape is pointed towards the substrate and may additionally be mounted on reciprocators moving both left–right and up–down. Typical applications include relatively simple elements such as chair legs.

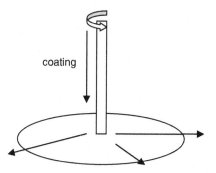

coating

Figure 15   Disc atomiser.

- *Disc atomisers*. A rotating disc driven by a turbine is mounted on vertical reciprocators such that the atomised coating is launched from the whole periphery of the disc. It is therefore necessary to move the target within the reciprocating droplet spray using an overhead conveyor that loops around the atomising disc (Fig. 15).

Because this movement has an omega (Ω) shape, the method is known as the 'omega loop'. Longer lengths are thus painted and the method is suitable for complex objects such as chairs. The transfer efficiency of such electro-assisted systems can reach values close to 90%.

2.2.7. Spray application of powder coatings

Powder coatings may be applied by fluidised bed or electrostatic techniques; however, as wood is unable to tolerate high temperatures only derived products, in particular MDF can be coated and electrostatic spray is the main method. Application on MDF substrates may be carried out on panels transported on horizontal conveyor belts or hanging vertically from overhead conveyors.

The advantages of horizontal application derive from the best uniformity and adhesion of the powder to the substrate. The main advantage of the vertical application is the opportunity to coat all the surfaces in one single pass. In both cases, the solid coating particles must be efficiently directed towards the substrate. The powder coating process comprises five major operational steps:

(1) Pre-heating of the substrate
(2) Powder application
(3) Powder melting and fusion
(4) Coating curing (by means of ovens, IR lamps or UV lamps)
(5) Cooling

Two different electrostatic application methods are used. They differ in the principle used to confer the electrical charge to the coating particles.

The 'Corona' systems use high-voltage generators and the 'Tribo' systems confer the electric charge by simple friction.

2.2.7.1. Corona systems   The corona system comprises four main components:

(1) The powder feeding system (tank and ducts)
(2) A gun for electrostatic application
(3) A high-voltage generator (around 100 kV)
(4) An air pump

The overspray is normally recycled by suitable recovery systems which separate air from powder and must not be contaminated when materials change. Corona systems are substantially similar to the electrostatic systems used for the application of liquid coatings with the electrode placed at the nozzle exit.

The powder particles directed by the airflow towards the substrate are charged by the ionising action of the electrode. Particle size must be carefully controlled as undersized particles become overcharged and will build up on sharp edges. Conversely, large oversize particles absorb too much charge and the overall coating thickness becomes variable. Free ions will also be present in the air stream and will be attracted to the earthed target where they may build up in recesses. The Faraday effect of back ionisation, or 'back spray', causes an uneven coating (orange peel effect) and will not penetrate recesses or evenly coat complex shapes. Increasing the velocity of the air may reduce the effect, but if it is too fast the powder will not have time to adhere.

2.2.7.2. Triboelectric or tribostatic systems   In this case the electric charge derives from the friction of the coating particles with the surface of complex ducts made of plastic. The intensity of the charge is usually positive if the plastic is PTFE, but also depends on the chemical nature of the powder. It is relatively low and so the repulsion among the particles is considerably reduced in comparison with the corona systems. The applied coats are then more uniform.

A positive effect deriving from the absence of free ions is the absence of back spray and less Faraday effect. The amount of coating material sprayed by a single nozzle is lower than the corona system. This drawback is overcome by the use of multi-nozzle guns.

As a consequence of the charging principle, the coating materials must be correctly selected and formulated; for example, epoxies and polyesters are naturally negative, whilst nylons charge positive. Besides the parameters described above, an important factor for the powder coating application is the conductibility of the substrate.

As in the case of solid wood, the panels used in combination with powder coating materials must have sufficient conductivity in order to properly connect to earth. Humidity and moisture content again play an important role. Although high humidity values promote the conductivity, the subsequent evaporation of water during the heating cycle can cause film faults (bubbles, whitening of the film, craters, etc.). Other possibilities for achieving necessary conductivity include the use of conductive primers, and making MDF more conductive through special additives.

3. PREPARATION OF THE SUBSTRATE

A unique feature of wood as a substrate is the propensity of wood fibres when wetted to undergo dimensional change and become more erect. The phenomenon of 'grain-raising' presents a major operational problem for high-quality wood finishing and will require sanding operations. In furniture coating, the nature of the sanding operation will be modified according to whether the furniture is to be coated as flat stock, or as a 3D-assembled article.

3.1. Sanding of the substrate

Sanding can be considered as the last operation in the manufacturing process of the uncoated product. It is carried out to remove the first wood layers, producing a smooth and uniform surface and also eliminating blemishes due to previous operations such as gluing. They are usually carried out by more passes just before the application of the first coat.

Sanding is carried out also on applied coats after drying and just before the application of the finishing coatings. In this case, the function of sanding is smooth any raised grain and also improve the inter-coat adhesion of the following coats.

The sanding processes can be:

Automatic. Totally automated sanding systems can be used especially in the case of furniture production from flat components. The sanding machines are equipped with two or three heads. The first head usually operates at right angles to the direction of panel conveyance. The direction of the following sanding heads (usually two) is the same as the panel direction. These two heads (typically belts) have different abrasive power the first being coarser than the second. These sanding machines are often equipped also with brushing systems removing finally the wood dust from the surface. It should be noted that wood dust presents a health and fire hazard and that proper precautions should be taken for extraction. Belt sanding systems are sometimes described as 'linishers' (Fig. 16).

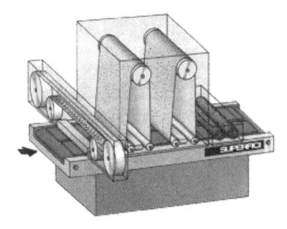

Figure 16   Automatic sanding machine (courtesy: Superfici spa).

Although belts are commonly used in sanding there are many other possible configurations including discs and bobbins. For high volume production machines may be configured to address a specific profile (profile sanders). Nylon brushes may be used for 3D surfaces such as assembled furniture; they are not as abrasive as sand but not suitable for hard surfaces.

Semi-automatic. In the case of wood floorings, for example, the sanding machines, equipped with powder suction devices, are manually moved by operators along the floors. In this case, two passes are generally carried out with abrasive materials of different sanding power. The abrasive materials are carried by belts, discs or drums.

Manual. Manual sanding is used especially with three-dimensional elements. It is carried out with brushes, sheets or pads. Flat sheets may be powered by vibrating heads, often using an orbital motion. Care must be taken to avoid swirl marks.

The abrasive materials used for sanding purposes can be different: sanding papers, brushing systems or metal wools.

3.2. Sanding papers

Sanding papers are constituted by a substrate (paper or tissue) to which the abrading material is adherent. The abrading material is made of mineral substances (e.g. silicates, aluminium oxides, etc.) of different dimensions. The abrading materials are mainly characterised by their dimensions and called 'grain'. In Table 2, a general sub-division of the sanding grades is listed with an indication of their common definition.

The 'grain value' represents the mesh per square inches of the sieve used for the separation of the abrading particles. The bigger this number,

Table 2 Sanding paper classification

Grain	Common definition
40–60–80	Very coarse
100–120–150	Coarse
180–200–240	Medium
280–320–360	Fine
400–500–600	Very fine
700–800–900	Ultra-fine

the greater is the number mesh inter-sections per surface unit and consequently the smaller the particles.

Other parameters characterising abrasive papers include:

- *Grain concentration*. A great grain concentration usually increases the sanding efficiency but, on the other side, a kneading effect can occur. The grain concentration shall be properly balanced to fulfil the specific exigencies.
- *Substrate*. The substrate (paper or tissue) confers properties as flexibility and mechanical resistance.
- *Glue*. The glue maintains the granules adherent to the substrate. It must be resistant to the heating effect deriving from the sanding process.

3.2.1. Metal wool

The use of metal wool imparts a milder effect in comparison with abrasive powders. Such materials are usually classified according to four classes depending on the diameter of the metal strings: 0, being the largest, to 0000.

There are also other proprietary materials (e.g. Scotch Brite®) which may combine various plastic, metal and abrasive materials in the form of pads or composite structures. These can be designed to minimise damage on less robust substrates. Such operations are found in window coating, especially after the application of the impregnating stain or the intermediate coating. In this case, a vigorous sanding operation is not appropriate due to the negligible coating thickness applied.

The use of iron wool is not recommended in combination with waterborne coatings. Residues of small iron particles, in combination with water, can lead to the formation of rust stains.

3.3. Brush sanding

Brush sanding is carried out on flat and moulded substrates by means of special machines equipped with cylinders or drums on which abrasive brushes of various types are carried. Often, two drums are configured to present the brushes at either side of the work piece (Fig. 17).

Figure 17 Automatic brushing systems (courtesy: Quickwood Esperia srl).

Brushing is carried out with different aims:

- In the case of moulded panels, the sanding operation can be carried out by using flexible lamellas constituted of abrasive paper. They are able to reach all the cavities of the substrate.
- Brushing with animal or vegetable fibres can be carried out after the application of stains promoting an adequate penetration in the wood pores and cavities removing also the coating excess. A particular

appearance 'tearing effect' can also be obtained in the reproduction of old fashion furniture.

- Brushing with the use of 'cleaning' materials is used to remove the powder formed by sanding of the substrate.
- Brushing can be carried out with cloth inserts to polish polyester or polyurethane topcoats. This operation is also carried out using waxes or polishing pastes.

3.4. Bleaching

Bleaching of certain wood species can be carried out to eliminate or reduce the natural colour of wood for different purposes:

- To eliminate possible isolated natural colour spots
- To harmonise the colour of different elements being assembled together
- To eliminate the natural colour of wood before the application of a stain
- To obtain light-coloured surfaces

Bleaching treatments involve also some practical disadvantages limiting their use:

- Bleaching represents an additional operation increasing the overall costs.
- The effectiveness of bleaching can be different depending on the wood species considered.
- Bleaching is frequently obtained by the use of strong oxidising substances. The possible residue of these chemical active species on wood can lead to the oxidation of some coating components. In particular in the case of polyurethane coatings, the oxidation of the aromatic group of the polyisocyanate causes discoloration (yellowing) of the coating film.
- The reactive substances used represent a potential risk for the safety of the operators.

The substance most frequently used as bleaching agent is hydrogen peroxide, which is a fairly strong oxidant. It is necessary to operate under basic conditions (low pH) in order to promote and accelerate the reaction. Typically treatments involve a combination of hydrogen peroxide with basic salts (as bicarbonates of sodium or ammonium), or with ammonia alone.

Other bleaching treatments based on oxidising or reducing substances are not very common in the wood sector. Occasionally, special treatments based on acids (e.g. oxalic or hydrofluoric) are used as they are effective to lighten local discoloration or to eliminate the effect of other contamination such as iron particles.

BIBLIOGRAPHY

CEFLA, Wood Coatings Manual, CFG CEFLA Finishing Group – via Bicocca 14-C – 40026 Imola (BO), Italy.

Speranza, A. (2003). La verniciatura del legno: preparazione delle superfici. *Pitture e Vernici* 11.

Perrin, H., and Zanazzi, R. (1982). Verniciatura a Spruzzo. *La rivista del colore, Milano.*

Karlsson, L. (2002). Triab UV powder paint line concept. *In* "Atti convegno CATAS sulla verniciatura a polveri", 6 March 2002.

Felici, N. J. (1978). Teoria dell'applicazione elettrostatica al rivestimento di superfici. *La rivista del colore, Milano – Verniciatura industriale* 11(118).

Svane, P., Hacq, Y. N., Gard, W., and Bulian, F. (1993). "Manuale – Rivestimento con polveri del legno e dei pannelli da esso derivati." CATAS.

Bulian, F. (2008). "Verniciare il legno." Hoepli, Milano. Christensen J. B., "Study on the Influence from Air Caps on Air Inclusion in Paint Films of High Viscosity Waterborne Paints. Paper 35. PRA Third International Woodcoatings Congress." October 2002, The Hague, The Netherlands.

CHAPTER 10

Film Formation: Drying and Curing

Contents			

Wood Coatings: Theory and Practice
DOI: 10.1016/B978-0-444-52840-7.00010-2

289

1. INTRODUCTION

During the process of film formation, all coatings undergo physical changes; in many cases, this will be accompanied by chemical changes such as cross-linking. For some coatings, film formation occurs at ambient temperatures, but for many industrial wood coating processes there is a need to provide additional energy in the form of heat or direct radiation. Operational aspects of drying and curing are described in the latter part of this section but first some of the basic concepts are introduced. Most coatings are liquid during the application stage with a viscosity suited to the application method as described in the preceding section. Powder coatings differ in that liquefaction occurs after application. In all cases, film formation requires at least one phase change which may be brought about by loss of volatiles, or in the case of powder solidification by 'freezing' on cooling. In either case, there may be further chemical reaction to bring about curing. The point at which a coating can be said to be 'dry' is not clear-cut and will depend on the test method used. In simple everyday terms, a coating is said to be 'touch dry' if it does not feel wet or sticky, and is unmarked by finger pressure. For this to be true, the coating needs to have attained a relatively high viscosity of around 10^3 Pa s. For finished work pieces to be stacked without blocking (i.e. sticking to each other), the viscosity must be much higher. Where a coating undergoes cross-linking to attain required properties, tests of resistance such as a solvent rub may be used. There are many other standardised test methods for assessing drying and curing.

Film formation is influenced by many factors including the environmental conditions such as temperature, humidity, air movement and radiation. Equally important is the nature of the polymer including molecular weight and morphology. Where a coating is capable of cross-linking then the presence of catalysts and accelerators will play a crucial role. Although there are many structural features of polymeric molecules that contribute to film formation, one of the most useful indicators of behaviour is the so-called 'glass transition temperature' (T_g).

Unlike small molecular compounds, which have a clearly defined melting point (T_m), synthetic polymers are at least partially amorphous. When the temperature is increased instead of a sharp increase in free volume at the melting point, there will be a more gradual change; this is a second rather than first-order transition. It is not a phase change but corresponds to the point at which there is an increase in the thermal expansion coefficient and arises from the ability of molecular segments to move relative to the molecule. T_g is often described as the point where the behaviour of a polymer changes from brittle to flexible. Strictly speaking this is not true, but it gives a qualitative idea of the significance of T_g in relation to the service conditions T_{env}. The T_g of a coating is not fixed but will change during film formation as plasticising solvent is lost, and if cross-linking occurs. The difference $(T_{env} - T_g)$ gives a useful guide to aspects of film formation since it is an indicator of the free volume through which small molecules such as solvent might move.

Mechanisms of drying are also strongly influenced by the technology route by which a particularly chemistry is delivered. This aspect was introduced in Chapter 3 and covers the physical form of the binder such as water-borne, solvent-borne, high solids and powder. In considering the operations of drying and curing, it is therefore useful to distinguish between:

- Deposition of dissolved polymer by solvent evaporation;
- Coalescence of polymeric particles after evaporation of water and other volatiles;
- Integration (breaking) of emulsified liquids on loss of water and other volatiles; and
- Melting and fusion of polymeric particles.

In each case, the initial stages of film formation may be followed by various cross-linking chemistries, and this will almost invariably be the case with emulsified polymers, which are usually of low molecular weight. UV and electron beam radiation-cured coatings may be 100% solids at the application stage, in which case film formation is solely brought about by the curing mechanism. However in some cases, it is operationally convenient to dilute such resins with solvent, or to emulsify them in water. In such cases, loss of volatile becomes the first stage of film formation.

Coatings are applied by many different methods, as already described in the previous chapter, and it could therefore be said that the very first stages of film formation are governed by the application method. Certainly, some volatiles will be lost during application. Once the coating is in contact with the target substrate, there will be various degrees of flow and in most cases good levelling is required. Levelling is driven by surface tension, which must overcome viscous resistance. For thicker films gravity can cause flow, usually as undesirable sagging. Because viscosity will generally increase during film formation, a point will be reached where both levelling and sagging are prevented. Achieving good appearance requires the right balance of viscosity and the external environment. For architectural (decorative) coatings, this must reflect the expected ambient conditions. Industrial wood coating uses many different configurations of equipment to bring about the necessary changes, as described in the latter part of this section.

2. FILM FORMATION BY EVAPORATION OF SOLVENT FROM SOLUTION: PHYSICAL DRYING

2.1. Non-aqueous solutions

Classical solvent-borne lacquers contain a relatively low concentration of high molecular resin (MWt $\sim$ 25,000) and will have a T_g around 80 °C. Typical examples of coatings belonging to this category are represented by cellulose nitrate products and others cellulose derivates (e.g. CAB). Cellulose nitrate binders present generally a glass transition temperature between 50 and 80 °C. They are usually plasticised by the addition of alkyds or specific additives.

Such materials would have a sufficiently high viscosity to be 'touch dry' at room temperature, but would be thermoplastic. Drying of a resin solution such as this will take place in two stages:

(1) Evaporation controlled
(2) Diffusion controlled

During the first stage, the evaporation rate is hardly, if at all, affected by the presence of the polymer and will depend on factors such as:

- Vapour pressure
- Temperature
- Polymer–solvent interactions
- Surface area of film
- Rate of airflow

In spray application, high surface area means that considerable solvent will be lost from the atomised droplets. Relative humidity has little effect on evaporation rate unless the solvent is water.

In the presence of solvent, the 'effective T_g' will be significantly lowered, but as solvent is lost T_g will increase and $(T-T_g)$ will decrease; thus, solvent loss becomes controlled by the diffusion rate through the free volume. As the effective T_g approaches the temperature at which the film is being formed, the rate of solvent loss will slow dramatically. For high T_g polymer solutions dried under ambient conditions, this means that significant (2–3%) amounts of solvent will become trapped, and may persist for years. Raising the drying temperature will reduce retained solvent, but to remove all traces the temperature of 'stoving' should be significantly higher than the actual T_g. During the diffusion stage of drying, the rate of solvent less will also be controlled by the distance it has to travel and will become inversely proportional to the square of the film thickness. The evaporation rate of single or mixed solvents is a poor predictor of solvent loss during the diffusion stage, and a fuller explanation involves solvent molecule size, and thermodynamic interactions. Evaporation is an endothermic process and will absorb heat from the substrate. Under some drying conditions, condensation of moisture on the coating surface will cause defects such as blooming.

2.2. High solids solvent-borne coatings

As noted in the opening chapters, solvent emissions to the atmosphere are subject to increasingly stringent regulations such as the European Solvent Emissions Directive and the use of low solids lacquers is in decline. Higher solids polymer solutions may meet the necessary regulations but raise some additional issues during the drying stage. It is usually found more difficult to prevent sagging after the application of high solids coatings in comparison with lower solids. This is partly because the films are thicker, but it also seems that less solvent is lost during application and that the transition from the evaporative to the diffusion stage occurs earlier [1]. One possibility is that the rate of change of effective T_g, for the type of low molecular weight resin ($\sim$15,000), which is necessary to raise solids content, is more rapid than for the higher MWt counterpart. Hence, the slower diffusion stage dominates solvent loss. Control of sagging in high solids coatings requires careful control of application parameters, such as atomisation for spraying, and temperature. Formulations may be modified with rheology modifiers that impart pseudo-plasticity or thixotropy.

2.3. Water-borne solutions

The use of water as a solvent presents additional complications from the perspective of drying:

- High latent heat
- Evaporation rate depends on relative humidity
- Hydrophobic interactions and associations
- Azeotropic effects with co-solvents

The high latent heat and heat of vaporisation of water make it slower drying from the perspective of industrial coating and will require more energy input. Unlike aliphatic solvents, the evaporation rate of water is slowed as humidity rises and will cease at 100% R.H. Many water-borne solutions, such as the water-reducible resins described earlier, contain co-solvents. Evaporation of components of the mixture will occur at different rates, depending on the ratios used and one or other component will become relatively enriched during drying. It is, however, possible that there will be a specific humidity where both water and the co-solvent evaporate at the same rate; this is known as the CRH (critical relative humidity) for the mixture [1]. Water can also form an azeotropic mixture with some solvents; this means the ratio of compounds in the liquid and vapour phase are the same and will remain constant during evaporation. Some co-solvents will therefore accelerate the evaporation of water. The influence of co-solvents is critical in controlling sagging, and popping (bubbles that leave a permanent crater) during the forced drying of water-borne coatings. Water-reducible resins are not truly dissolved in water but are present as swollen aggregates; this can cause some partitioning of solvents between the resin and solvent phase and in turn this will affect the CRH. Such resins also show abnormal viscosity effects when diluted with water, and on evaporation, because the water:solvent ratio is changing which in turn changes the association of ionic molecules.

2.4. Water-borne emulsions

Water-borne emulsions that are liquid resins emulsified into water and not a polymeric particle as described below, they find increasing application in wood coatings as a means to lower VOC. As noted earlier alkyds, polyol (for isocyanate cure) and epoxy resins have all been emulsified. In all cases, the first stage of drying is governed by evaporative factors, but at some stage there will be a phase inversion and the liquid resin will become the continuous phase through which water must migrate.

2.4.1. Water-borne dispersions – Latexes

Film formation of water-borne dispersions is a multi-stage process that involves the loss of volatile material, particularly water, and the fusing together of dispersed particles to form a more or less coherent film. The fusion stage is usually called coalescence (from the Latin word $co + alescere$ = to join together) and is seldom complete under ambient conditions. During the process of coalescence, T_g plays an important role and will be lowered by the presence of water, or deliberately added co-solvent known then as coalescing aids.

Film formation of water-borne dispersions has been extensively studied [2] and is usually divided into four stages (Fig. 1):

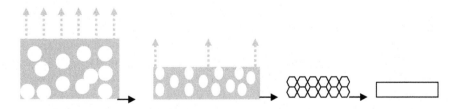

Figure 1   Stages in the coalescence of latex particles (Schematic)

(1) Stage 1 – evaporation
(2) Stage 2 – agglomeration and deformation of particles
(3) Stage 3 – integration and coalescence
(4) Stage 4 – inter-diffusion

In some accounts, the final stage is omitted or combined with the third, others concentrate on only two main stages. However, it is possible to separate at least three mechanisms, which have important practical consequences in drying and development of film properties and these are summarised here. Short accounts such as this are necessarily simplified and in the drying of a large area the stages will overlap. The transition from stage 1 to 2 occurs on an advancing front, which may break up into 'islands' of activity.

2.4.1.1. Stage 1 – Evaporation   The evaporation stage of water-borne dispersions lasts until the particles start to become in close contact (~60–70% phase volume) and is governed by the factors described above including temperature, humidity, rate of airflow, evaporative cooling and the presence of any 'co-solvent'. By the end of this phase, most of the water has been lost and particles will be subject to repulsion forces according to the charge or steric stabilisation of the particles; these are not usually strong enough to resist capillary forces.

2.4.1.2. Stage 2 – Agglomeration and deformation   Once the particles are in contact, the pathway for further water loss becomes more tortuous and the overall rate of evaporation decreases. With soft particles (or at higher temperatures) particle deformation occurs, and interfacial tension closes capillary channels. The lowest temperature at which deformation, followed by coalescence, can occur is known as the MFFT (minimum film-forming temperature). MFFT can be measured by placing a coated glass panel on a temperature gradient bar, and noting the transition from a clear to a cloudy layer. Commercial MFFT bars are available for this purpose [3].
    The details of the process are complex and will differ between polymer types; theories of fusion include:

• Dry sintering
• Wet sintering
• Surface compression
• Inter-particle cohesion

Stewart [2] gives a useful review of these theories.

2.4.1.3. Stages 3 and 4 – Integration and diffusion   This final stage of film formation involves the loss of residual water, initially through any remaining channels but ultimately by diffusion. However, water-soluble material between the particles may provide diffusion pathways. For many practical types of latex, the permeability is higher than would be expected from the same polymer cast from solution. It is important from the perspective of both practical performance and testing to recognise that film properties will be very dependent on their recent history. Raising the temperature may increase integration (annealing) and water can leach out surfactants and other soluble materials. Some polymeric particles show further diffusion (autohesion) but with many cast films and the right techniques; it remains possible to discern polyhedral particle boundaries even in mature films.

As with solvent-borne solutions, the effective T_g plays an important role in film formation, and the MFFT (minimum film formation temperature) will be strongly influenced by $(T-T_g)$. The effective T_g will be modified by any heterogeneity within the polymer particle, and the plasticising effect of solvents including water itself. With polymeric particles, there is a conflict between the need for a low T_g to promote film integration and a higher T_g to enable good mechanical properties. The possible strategies for overcoming this problem include:

• Film forming at higher temperatures
• Addition of a temporary but still volatile plasticiser
• Use of co-monomers plasticised by water itself
• Mixtures of hard and soft particles
• Enabling a soft outer shell to a hard particle (core–shell morphologies)
• Using a low T_g polymer that can subsequently be cross-linked

Higher temperature film forming is an option for industrial coating but not for architectural paints, which must dry, and cure under ambient conditions. The addition of a temporary plasticiser, known usually as a coalescing aid, is still widely used but under some pressure from solvent regulations. The choice of coalescent, or blend, is system specific and depends upon partitioning of the solvent between water and polymer phases. Examples include alcohols, ether alcohols and their esters. Some co-monomers, for example, VeoVa (vinyl esters of versatic acid) are plasticised by water, but most are not. Polyurethane dispersions (PUDs) have a water-swollen morphology which aids film formation.

Other approaches to the problem of lowering the MFFT without sacrificing film properties have included polymer blending and the use of sequential emulsion polymerisation to obtain a 'gradient morphology'. The blending of a high and low T_g latex can certainly give intermediate properties but percolation effects mitigate against getting the ideal balance this way. Latex blending will also introduce interactions between the two lattices and other ingredients, which complicate formulating and also increase operational complexity. Alteration of latex particle morphology can be achieved through control of both kinetic and thermodynamic aspects of emulsion polymerisation. In principle, a 'core–shell' morphology

offers an additional degree of control between the process of film formation and subsequent dry film properties.

Low T_g thermosetting latexes are a possibility for industrial coating
but in many cases require a two-component approach. For example, urea
and melamine formaldehyde are used for hydroxyl and carboxy functional polymers. Carbodiimides and poly-functional aziridines are also
used.

3. POWDER COATINGS

Both thermoplastic and thermosetting powder coatings are used, and in
both types the fist stage of film formation is to melt the powder at a
sufficiently high temperature to allow flow, and then to form a film by
cooling, and 'freezing'. Once again T_g is a critical parameter requiring
compromises. If the T_g is too low the particles will sinter together on
storage, and if the T_g is high then a very high temperature is required for
fusion. For a polymer to reach a viscosity at which it can flow, the fusion
(i.e. stoving or baking temperature) must be at least 70 °C above the T_g of
the uncured powder. Since the latter is typically at least 55 °C, the stoving
temperature needs to be above 125 °C, with clear implications for wood
coatings. Hence at present powder coating tends to be restricted to MDF.

The driving force for fusion of powders and subsequent levelling has
been less studied than for latex coalescence, but is likely to be driven by
surface tension forces and surface tension gradients. Pigmentation of
powder coatings causes a disproportionate increase in melt viscosity
since there is no possibility of using a volatile component to reduce
viscosity. Many powder coatings contain small amounts of benzoin
($<1\%$), which helps to prevent pinholes, and allows some degassing.
Benzoin has a T_g of around 134 °C, and it may act as a plasticiser, thus
lowering the effective T_g.

4. FILM FORMATION AND CHEMICAL CROSS-LINKING

4.1. Thermosetting resins

Film formation by solvent evaporation alone is applicable to thermoplastic polymers but carries the penalty of a high solvent content in order to
dissolve a relatively high MWt polymer. The alternative is used a lower
MWt polymer followed by a chemical reaction to raise the MWt. Such
resins and polymers are usually described as 'convertible', or thermosetting. Even higher MWt resins and dispersions may benefit from additional cross-linking for the more demanding applications.

All chemical cross-linking reactions raise the same generic problem for coatings technology, namely how to prevent premature reaction and ensure adequate pot life. This may be achieved in various ways, for example, by protecting the bulk polymer from one or more of the reactive components. Thus, air-drying alkyds are projected from oxygen in storage. Another possibility is to separate components as in a two-pack (2K) coating. It is also common to design coatings with low reactivity under ambient conditions, and then raise the reaction rate by an increase in thermal energy (stoving) or other radiation (UV curing).

In general, chemical reaction rates are governed by the Arrhenius relationship and will double for each 10 °C increase in temperature. However, the behaviour of polymeric reactants is complex and the simple relationship is not always followed; there may also be sequential or reverse reactions, as in the use of 'blocking' agents. Although the reaction rates of coatings designed for stoving and other means will be considerably slower under ambient conditions, they are still capable of slow reaction and the pot life may be limited, storage at low temperatures may be advisable.

The effect of T_g should also be taken into account for thermosetting reactions. Initially, the T_g will be low and diffusion of reactants takes place readily. As the T_g rises, the diffusion rate of reactants will slow unless the stoving temperature is well above the new T_g. For reactions under ambient conditions, it is likely that the T_g will rise above the ambient temperature and the reaction rate will be considerably slowed. Further reaction might continue for months or even years. Resistance to dissolution is a quick way of testing whether the reaction has proceeded. In stoving reactions the oven temperature is usually well above the T_g and diffusion of reactants is not limited. This can be more of a problem for powder coatings since the initial T_g is already high to prevent the powder sintering on storage.

5. REPRESENTATIVE CURING TECHNOLOGIES

The means by which a polymer is converted from liquid to solid form has significant operational implications with options for mixing components and in some cases curing at elevated temperatures. Wood is more constrained than metal substrates in this latter respect.

5.1. One-component physically drying coatings – Lacquers

Here, the drying mechanism is regulated only by the solvent evaporation as no chemical reaction is involved. The MWt of the film former must be sufficient to allow the formation of a coating film with acceptable properties. Typically, this requires a linear morphology to be soluble in a suitable organic solvent (or blend).

Typical examples of coatings belonging to this category are represented by the cellulose nitrate products and others cellulose derivates (e.g. CAB). Cellulose nitrate binders generally have a glass transition temperature around 50 °C. They are plasticised by the addition of alkyds or specific plasticisers which may be fully or partially compatible such as phthalate and phosphate esters.

The presence of high molecular weight film formers is usually associated to the use of considerable amounts of solvent in order to adjust the viscosity to the values required by the application systems.

The main factors affecting the drying process of such products were described in Section 2.1.

5.2. One-component chemically drying coatings (ambient temperatures)

The most representative coating belonging to this category is illustrated by drying oils and alkyds reacting with oxygen.

The drying process is influenced by the solvent evaporation together with some parameters deriving from the specific chemical reactions involved in autoxidation. Auto-oxidative curing is an example of a step growth homolytic reaction; it has been extensively studied [4]; the mechanism is complex with differences that depend on the conjugation of the oil component. During the curing process free radicals are formed that react with oxygen to form peroxides and then hydroperoxides. These in turn generate further radicals which undergo combination and other reactions which build up the molecular weight and the film becomes cross-linked. The process is catalysed by various 'driers' which perform specific functions in the overall process. Driers are thus often described as 'surface' or 'top' driers and include cobalt and manganese salts (e.g. of octanoic acid). 'Through driers' include zirconium and catalyse reactions within the film. They may be supplemented by other materials which have synergistic effects. Reactions do not cease when the film is dry and can cause subsequent problems such as yellowing, and in the longer-term embrittlement.

Another type of one-component chemically drying coating can be obtained by 'blocking' a reactive component such as an isocyanate. However, as elevated temperatures are usually necessary to unblock the reactant, these might also be categorised as stoving products. An exception is the case of hydroxyl-terminated polyesters which have been reacted with an excess of diisocyanate. These can be activated by exposure to water (moisture curing) which creates amines that in turn react to form urea cross-links. Clearly, such materials must be protected from water during storage; the reaction generates carbon dioxide bubbles which may become trapped if the viscosity is too high.

5.3. Multi-pack chemically drying coatings

The following examples can be considered as representative coatings belonging of this category:

- Two-pack polyurethanes
- Unsaturated polyesters
- Amino coatings (acid curing)

The drying process is greatly influenced by the reactivity of the components and the presence of catalysts. Two-pack (two component = 2K) material such as the polyurethanes derived from isocyanate/polyol reactions requires an accurately metered mixing ratio, and will have a limited pot life. Epoxy coatings cured by the addition of polyamide have similar constraints but are hardly used in wood coating.

Unsaturated polyesters and acid-catalysed amino resins have both reactants in the one container and only the catalyst is kept separate. They will have a limited pot life after activation but the mixing ratio is rather less critical.

5.4. Stoving coatings

Such coatings require heat to bring about the cross-linking reaction and at high temperatures can be very fast. High-temperature stoving is used in industrial metal finishing (120–180 °C), for example, for various amino resins, but this is too high for wood. However as described below, various types of forced drying at somewhat lower temperatures are used in industrial wood coating to speed both solvent loss and reaction rates. Only some heat-curing powder coatings, used with MDF, can be properly considered as part of this group.

5.5. Photo-curing coating

Photo-curing products also belong to the category of chemically drying coatings. The aim is again to obtain high molecular weight cross-linked structures starting from more simple molecules. The process starts from linear polymers in combination with reactive monomers.

The chemical reaction takes place in three steps:

(1) *Initiation.* The production of active species (free radicals) is initiated by means of ultraviolet radiation acting on photo-initiators, or by the use of electron beams (EB) promoting the formation of radicals directly on the binder.

(2) *Propagation*. Radicals so formed are able to react with the double bonds present in the chemical structure of polymers and monomers. During this phase, the polymer structure rapidly grows.
(3) *Termination*. When a radical reacts with another, the chain reaction is terminated. It shall also be considered that the action of such highly reactive species is progressively slowed down to a negligible degree by the reduced mobility in consequence to the growing of the molecular dimensions of the system.

The factors affecting such process derive mainly from the equipment used and in particular by lamps (number, type and configuration) in the case of UV systems. For three-dimensional objects, the orientation of the surfaces with respect to the lamps will also influence the curing time as will any shadowing. Free radical UV curing is fast but stops once the irradiation ceases, the rapid gelation traps some radicals and it is difficult to get 100% reaction of all double bonds. Degree of cure is lower for pigmented products which may absorb UV. Residual initiators and their by-products, such as benzaldehyde or benzyl alcohol, can give rise to lingering odours in the coatings.

In the case of EB technology, the initiation phase takes place without the use of photo-initiators. The radical species are produced directly on polymers and monomers by the high-energy electrons involved in such technique. Formulations are generally cheaper but the equipment cost is much higher.

6. INDUSTRIAL DRYING PROCESSES

Architectural coatings are of necessity carried out under ambient conditions and this presents some constraints when weather conditions are poor. High humidity and cool conditions are particularly problematical for decorative water-borne coatings. For industrial coating drying and coating conditions can, and must, be controlled. The equipment used will depend on the market sector and can range from simple forced drying to sophisticated automated plants. Some operational factors are outlined below.

6.1. Heating sources

The most widely used systems for the drying of all coatings types are based on heat transmission. Heat improves the evaporation rate of solvents, including water, and accelerates the possible chemical reactions involved. Heat can be transferred to the coating by simple convection or by irradiation.

Table 1 Radiation type and physical process involved

Radiation type	Physical process involved
IR	Molecular vibrations
Microwave	Molecular rotations

In the second case, infrared (IR) or microwave radiation sources are used, causing different effects on the molecular structures of the coatings (Table 1).

6.1.1. Heat convection

Convection ovens or tunnels are equipped with heating devices able to direct, a flux of heated air towards the coated substrate. This is achieved directly by means of electrical resistance heaters, or indirectly (e.g. using heat exchangers). Heat is also transferred to the liquid coating by convection (air movement), equally importantly volatile materials are removed. With IR, the upper surface of opaque coatings will be the first to warm up before heat is conducted to the lower parts. There is therefore a risk of inhomogeneous drying, leading to defects such as solvent retention or bubbles (pinholes). Lower temperatures and longer drying times are advisable when high thicknesses are applied. Often, a better alternative is to apply multi-layers with a flash-off period between.

Ventilated ovens are the most widely used systems for the drying of coatings in the wood sector because of their suitability for elements of different shape and dimensions. Beside the dimensions, the other significant parameters of such systems are: temperature and ventilation. Temperatures are relatively low (usually between 20 and 40 °C) considering that the substrates (wood or wood derivates) are particularly sensitive to such parameters. For water-borne coatings, the relative humidity must be below 75%. Air velocity will be around 0.2 m s^{-1} at the surface. Ventilated ovens are suitable for drying of most coating types, sometimes in combination with other systems. Convection tunnels are sometimes used without external heating, relying upon air movement to remove solvents; however, such 'cold drying' does require a storage period before assembly and stacking becomes possible.

Heated convection drying has many advantages for large 3D objects, though parts of the object may be partially shielded. However, it is not a very efficient method of heat transfer and there is energy wastage. Sawdust may be used as a source of fuel for incineration and heat exchange.

6.1.2. Infrared lamps

IR lamps emit radiation that, when absorbed by matter, is directly transformed into heat. Absorption depends on the chemical composition and surface colour of the coating. There are many different types of IR lamps available. They differ in the wavelength emission, which derives from the temperature of the lamp. The higher the temperature the shorter the wavelength produced. Shorter wavelengths penetrate transparent coating layers more effectively and directly heat the substrate. In principle, drying proceeds from the substrate, reducing solvent 'trapping' and the other disadvantages typical of convection systems (formation of blisters or surface skin). Nonetheless, too rapid application can still cause blisters to form.

Short and medium IR wavelengths emitting lamps are mostly used when a relatively strong heating effect is required. This would be typical of water-borne coatings due to their high latent heat of evaporation, and also for powder coatings, as they need to be both melted and cured. Longer wavelength lamps are more suitable when a smaller heating effect is required. However, their effect is mainly due to a convection process as described above. Convection is less sensitive to the colour of the coating or substrate.

The main advantages of infrared systems are:

• Low-energy consumption as they mostly heat the object, without heating the surroundings
• Efficiency towards all coatings type
• Easy control
• Low cost

A disadvantage of IR lamps is in controlling the distance from source to coated object and the need for a direct 'line of sight' to the surface. To overcome this limitation, hybrid systems using both infrared and convection technology have been developed. As the infrared lamps generate heat, blowers are used to circulate heat inside the oven, also drying the shielded parts and improving uniformity.

6.1.3. Infrared plates

Another type of infrared system is represented by catalytic gas IR plates. Such plates use a combustible gas (e.g. methane) instead of electricity to generate the infrared radiation. The gas is oxidised by the catalytic system at a relative low temperature. As the catalysis reduces the activation energy of the oxidising reaction, the plates' energy consumption is relatively low.

Like electric infrared lamps, the gas catalytic plates may be combined with air blowers to circulate the heat that is produced. The advantage of catalytic infrared plates mainly derives from their dimensions, allowing the drying of large elements (e.g. window frames) and reduced energy consumption.

6.1.4. Microwave sources

Microwaves are electromagnetic radiation with a wavelength shorter than the IR, and able to resonate the chemical bonds of water molecules. A molecular vibration is produced in which the radiant energy is absorbed and converted into heat. Microwaves systems can be used only in combination with water-borne coatings. This principle makes possible the simultaneous heating of the entire liquid layer applied onto the substrate leading to drying from the inside to outside, and therefore less prone to blistering. However, any variations in film thickness will give uneven heating.

Microwaves are dangerous to human health and so all the drying area must be adequately shielded. Microwaves which act primarily on water molecules are ineffective towards organic substances like coalescing solvent. Coalescents should therefore be eliminated by other heating systems (e.g. ventilated ovens or IR systems) following the microwaves drying. Microwave drying is sometimes used only to evaporate water, and the coating is subsequently fully cured by UV methods.

6.2. Radiation-curing systems

The basic principles of UV photo-curing have been presented in the previous chapters. The advantages of such technology include greater productivity and low VOC emission of correctly formulated coating materials. Very durable finishes are produced with good resistance to mechanical damage at an early stage. However, capital investment is higher, some reactive diluents present skin irritant hazards and the technology requires adaptation for 3D rather than flat panels. Cooling of the lamps may be necessary.

UV lamps are characterised by the wavelength and intensity of the radiation emitted. The emission range depends on the type of source used for the construction of the lamp. The most commonly used are based on the presence of two metals (mercury and gallium) in their vapour state with tungsten electrodes. Other lamps are also available containing additional elements (e.g. indium, iron) added in small amounts. These lamps are called 'doped' lamps. Small amounts of halide can shift the radiation output towards the visible spectrum. The lamp power depends on the gas pressure inside the lamp. It is also possible to use electrode-less lamps in which microwave power is used to excite the mercury atoms. The size of these lamps is restricted to 250 mm and they are positioned in groups. Intensity during the curing process can be controlled indirectly by monitoring the cure of the irradiated coating. Alternatively, control systems and tools such as radiometers can be used to monitor and possibly adjust the output or the speed of the conveyor belt. The performance of lamps does change over a period of use, and nominally similar individual lamps may vary in their output. The amount of ventilation may also have observable effects on the operating behaviour of lamps. For this reason, some monitoring of performance is advisable.

Lamps can be broadly divided into three categories:

(1) Low intensity (or pressure) = up to 10 W cm^{-1}
(2) Medium intensity = around 30–50 W cm^{-1}
(3) High intensity = 80–120 W cm^{-1}

Selection of the appropriate lamp depends on the nature of the formulation. Some photo-initiators are sensitive towards specific wavelength. The opacity of the coating is also important. Mercury lamps show a maximum emission (energy peak) at 366 nm and are appropriate for clear coatings. Gallium lamps have two main emission peaks at 410 and 420 nm (Figs 2 and 3).

This particular emission band makes possible the curing of paints containing titanium dioxide (TiO_2). The letter pigment is of major importance in white and pastel coloured coatings and at the correct particle size strongly scatters visible and some UV light. However, above 400 nm rutile and 380 nm anatase, TiO_2 becomes progressively more transparent to UV radiation. Photo-initiators with sulphur or nitrogen containing groups absorb more towards 400 nm and are used to cure pigmented coatings; they may, however, impart some colour. A UV-curing unit is generally configured with one or more lamps. Typically, the uncured coating passes beneath a lamp system on a conveyor belt. Radiation dosage is controlled by both the power of the lamps and the speed of the belt. Low-energy lamps are may be used to pre-gel the coating before curing is carried out by medium or high-energy lamps. This could require a combination of different units (e.g. gallium + indium). The first promotes the internal cross-linking of paint while the second is more effective on the coating surface increasing its hardness. Operational aspects to be considered are heat and ozone production. Radiation from the lamps may be focussed onto specific areas using parabolic reflectors, or de-focussed onto a broader coating surface by elliptical or intermediate reflectors. Reflectors are typically anodised aluminium and will reflect both UV and IR radiation. The fortuitous heat may be beneficial for solvent removal, but can present overheating problems requiring ventilation. Another possibility for heat sensitive substrates is to use dichroic reflectors which help transmit heat away, while reflecting UV towards the target (Fig. 4).

Pulsed xenon lamps have a high power output but because of their intermittent emission, they suffer less substrate heating. Characteristically,

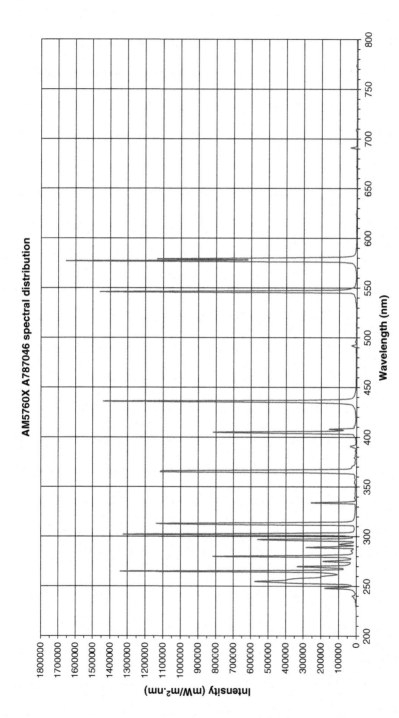

Figure 2 Mercury lamp – spectral distribution.

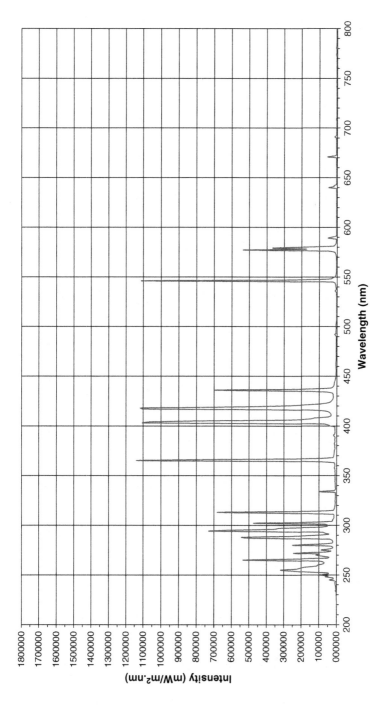

Figure 3 Gallium lamp – spectral distribution.

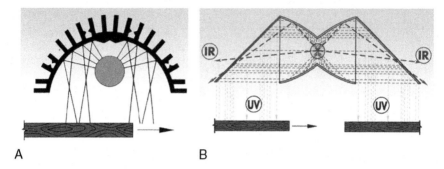

Figure 4   Reflectors: (A) parabolic and (B) dichroic (courtesy: Giardina spa).

most lamps will generate ozone due to the interaction between UV light and oxygen. Ozone is a dangerous gas for the safety human beings being produced by the interaction of very low wavelengths of the UV range with the atmospheric oxygen; it can be reduced by the adoption of special glasses for the lamp construction which are able to absorb some of the radiation but it is still necessary to remove ozone by good extraction systems (to below 0.1 ppm). Where air inhibition of the surface is a problem, systems may be purged with nitrogen and this has the added benefit of preventing ozone formation. UV curing is ideally used on flat surfaces, but lamps may be placed around objects, and some use of robotics can extend applications into some 3D areas.

To summarise the factors on which the UV-drying systems are optimised are:

- Type and number of lamps
- Distance of the lamps from the substrate
- Adoption of shield for the irradiation
- Configuration of reflectors
- Transport speeds
- Ventilation and extraction
- Pre-heating (for water-borne coatings)

6.3. Electron beam

Accelerated electrons (beta-rays) can trigger photo-chemical reactions (poly-addition) directly, without the need for photo-initiators. Reactions are complete in less than a second and allow very high coating speeds and deep penetration, even into pigmented coatings. Penetration is proportional to the beam voltage, and rate of cure to the beam current. EB-curing equipment generates electrons in a vacuum chamber, rather like a cathode-ray tube, and accelerated and transmitted through a cooled titanium foil window. The gap between the emitter and the coating is purged with nitrogen. Energy consumption is low but capital costs are higher. Retardation of electrons generates short-wave X-rays and the equipment must be well screened by lead, or thick concrete protection. The main advantages of the EB technology can be resumed as follows:

- Rapid production speed
- Photo-initiators are not needed
- The curing efficiency is almost independent of the coating thickness
- It is possible to cure opaque paints of any colour
- The gloss appearance can be controlled and adjusted

However, electron beam-curing systems are rather complex and expensive furthermore EB technology can only be used for flat surfaces. It is therefore less common than UV in the wood sector; applications include flat doors with opaque finishes.

7. INDUSTRIAL PLANTS

Industrial drying plants can be configured to be in a continuous or batch mode. In the first case, the units are transported directly to the drying area by the same transport system as is used for the application process. In the case of the batch systems, the two processes are separated. In batch mode, it is common for ovens to be fitted with doors whereas continuous systems make more use of tunnel ovens. The plants can be equipped with one or more of the drying and curing systems as described above. In the following paragraphs, the most common industrial set-ups are briefly introduced.

7.1. Horizontal tunnels

Horizontal tunnels are automatic continuous drying systems used for flat substrates comprising a conveyor belt passing through an insulated oven. Horizontal ovens can be equipped with different drying systems depending on the coating material to be cured. In the case of water- or solvent-based stains, the tunnel can operate by the simple use of hot air (in which airflow is in the reverse direction to the workflow movement). Heat may be generated indirectly via a heat exchanger and either directly or indirectly with IR lamps (Fig. 5).

Air movement is sometimes induced through nozzles pumping air at very high speeds. These tunnels are called percussion tunnels and are particularly efficient for the fast drying of water-borne stains (Fig. 6).

In the case of polyurethane, polyesters and other solvent-based coatings, drying tunnels are mainly based on low heating levels provided by

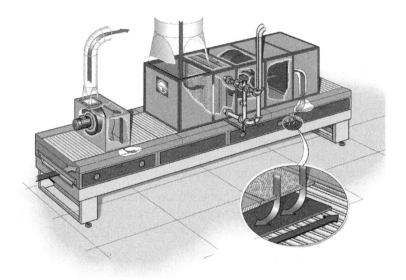

Figure 5   Heating tunnel – convection (courtesy Giardina spa).

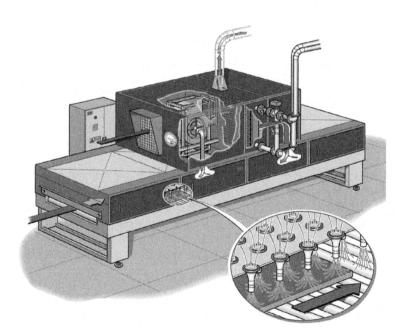

Figure 6   Heating tunnel – percussion (courtesy Giardina spa).

convection systems. For the drying of water-borne coatings, the presence of IR lamps can provide a more efficient result.

Such tunnels are generally divided into three parts:

(1) *Flash-off area*. This portion of the tunnel is the one immediately after the application. During this phase, the substrate is not heated because the liquid coating requires time for adequately levelling and the migration of some components (e.g. waxes) in order to confer desired properties to the surface. The flash-off phase is also important in order to allow any air bubbles to escape from the liquid film. During this phase, a considerable amount of the solvents (if present) evaporate from the substrate.

(2) *Heating area*. During this phase, the liquid coating becomes a solid film in consequence of the evaporation of the majority of the residual solvents and/or for the simultaneous chemical reactions involved.

(3) *Cooling area*. Panels and other units must be cooled down before they are handled in order to prevent damage to the soft coating. Units are usually stacked after the exit from the tunnels where they are vulnerable to unwanted adhesion (known as 'blocking'). Tunnels ovens specifically for water-borne coatings may be equipped with microwave generators (Fig. 7).

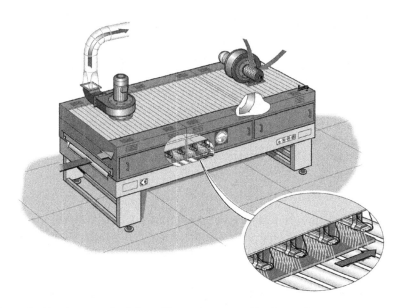

Figure 7   Heating tunnel – infrared lamps (courtesy Giardina spa).

In tunnels for UV curing, the following additional considerations apply:

(1) If the coating material contains negligible amounts of organic solvents (being effectively 100% solid content), the tunnel starts with the so-called gelling phase. In this portion of the tunnel, low-intensity lamps initiate the cross-linking reaction. This phase is important to allow a gradual reaction improving: homogeneity, adhesion and migration towards the surfaces of the specific additives. The last part of the tunnel equipped with high-intensity lamps completes the reaction.
(2) If the coating material contains significant amounts of organic solvents, the tunnel must be provided with an initial heating phase (usually hot air) just before the UV lamps. This phase is important in order to allow the solvents to escape from the coating. After this initial stage, the tunnel usually continues with the two parts (gelling and curing) described above.
(3) For UV curing of water-borne coatings, the initial removal of water is more critical. The presence of even slight amounts of water in the cured film usually causes whitening, or a significant loss of transparency. This is especially true where accumulation of moisture is possible (e.g. in cavities or corners). Usually, a combination of hot air and IR lamps (or microwaves systems) is used.

The general advantages of tunnel systems derive from their productivity, automation and insulation from the possible deposition of dust on the surfaces during the drying phase. The tunnel length, and consequently its speed are connected with the application phase. Dimensions depend on various factors: including coating materials used and the geometry and size (thermal mass) of the work piece. Some tunnels are designed in a U shape to reduce the footprint occupied by the plant.

It is good practice to investigate the energy balance of convection and other ovens with a view to energy savings. Such a study should analyse energy losses at all stages of the process. The chemistry of curing has a major impact on energy requirements and can be reduced by higher solids systems and lower temperature-curing chemistries.

7.2. Vertical ovens

Vertical ovens can be considered as continuous drying systems for flat or moulded substrates which are configured vertically rather than horizontally. They have been developed to avoid the problem of long drying times necessary for some coatings where the consequence would be a productivity reduction or an excessive length of tunnel.

Vertical ovens comprise stacked chambers where panels automatically move, from one chamber to the other, on shelves. The three phases previously mentioned above (flash-off, drying and cooling) are internally present.

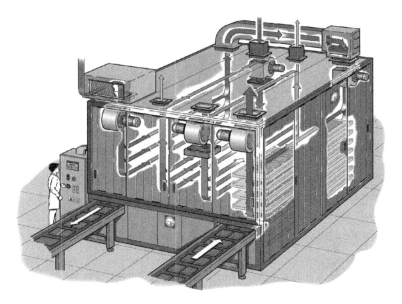

Figure 8   Vertical oven (courtesy Giardina spa).

The loading stage, on conveyor belts, is usually automatic with synchronisation systems based on photo-cells. After entering the oven, panels transport is based on ascendant and descendant movements. Drying is carried out by hot air circulation. Some vertical ovens, called 'lungs', comprise a single chamber being used just for the flash-off phase before the real curing reaction under UV lamps (Fig. 8).

7.3. Multi-level ovens

Multi-level ovens have been largely superseded the vertical oven and are not much used. They were developed to overcome the problem of long work pieces especially when high thickness coatings, requiring long drying times, are applied. The basic concept is to reduce the speed of the conveyor belt in order to maintain the substrate inside the oven for the required time. The reduction of the speed allows the use of smaller ovens. The presence of more stacked ovens on different levels compensates for the reduction in productivity. Also, these systems are suitable for flat or moulded panels. The length of such ovens is longer than the vertical ones.

7.4. Tunnel ovens for multi-layer racks

These tunnels, using hot air circulation, represent a sort of parallel drying line to the application one. The coated elements (flat or moulded elements) are initially placed on multi-layer trolleys or carousels that, starting from the application area, run through the drying tunnel. As with the other configurations, all three drying stages are usually present. After the final discharge, the trolleys return automatically to the application area.

7.5. Ovens for three-dimensional elements

The dimensions of such ovens must be capable to contain fully assembled products such as chairs. The systems can be automatic. The furniture elements run inside the oven hanged to a suitable overhead conveyor immediately after the application phase. Such ovens can be considered as continuous drying systems.

7.6. Drying chambers

These systems are simple hot chambers where flat panels, placed on suitable racks, or assembled furniture are placed. They are especially suitable for water-borne coatings. It is advantageous to use dehumidifiers as a more homogeneous and quicker result can be reached.

Drying chambers are classified as batch drying systems as the coated substrate will be introduced and discharged separately from the application line.

REFERENCES

[1] Wicks, Z. W., Jones, F. N., and Pappas, S. P. (1999). "Organic Coatings: Science and Technology", 2nd edn, 630 pp. John Wiley & Sons, Chichester (ISBN 0-471-24507-0).
[2] Stewart, P. A., et al. An overview of polymer latex film formation and properties. Advances in Colloid and Interface Science Volume 86, Issue 3, 28 July 2000, Pages 195–267.
[3] Sheen Instruments (Hartest) (http://www.sheeninstruments.com/products/process-control/mfft.htm).
[4] Carr, C., Dring, I. S., and Falla, N. A. R. (1990). Studies of the curing of soya oil/alkyd films containing metal driers. In "Proceedings of the 16th International Conference in Organic Coatings Technology", Athens, pp. 99–109.

BIBLIOGRAPHY

Wicks, Z. W. (1986). Film formation, Philadelphia, PA, Federation of Societies for Coatings Technology.
CEFLA, Wood Coatings Manual, CFG CEFLA Finishing Goup – via Bicocca 14-C – 40026 Imola (BO), Italy.

Printed in the United States
By Bookmasters